Evaluating Mineral Projects:

APPLICATIONS AND MISCONCEPTIONS

THOMAS F. TORRIES

Published by

SME

Society for Mining, Metallurgy & Exploration

Society for Mining, Metallurgy, and Exploration, Inc. (SME)
12999 East Adam Aircraft Circle
Englewood, Colorado 80112
(303) 948-4200 / (800) 763-3132
www.smenet.org

SME advances the worldwide mining and minerals community through information exchange and professional development. With members in more than 70 countries, SME is the world's largest association of mining and minerals professionals.

ISBN 10: 0-87335-159-2
ISBN 13: 978-0-87335-159-1
Ebook: 978-0-87335-279-6

Contents

List of Figures

List of Tables

Preface

Project evaluation is the process of identifying the economic feasibility of a project that requires a capital investment and making the investment decision. The subjects covered in this book apply to all types of projects for which investment decisions must be made. However, the focus of this book is on mineral projects, which often present unique evaluation difficulties. Complex mineral projects often have many uncertainties caused by the very nature of geologic reserve estimations, severe problems in forecasting commodity prices and production costs, long evaluation periods during which economic and technologic conditions can change, uncertain regulatory and environmental costs, and, in many cases, long project lives. Much care and perhaps multiple evaluation methods are required to obtain results on which to base mineral investment decisions.

Other, more general factors are also important. An evaluation method must anticipate both the future course of the project's operation and the future behavior of the market in order to yield accurate results. In addition—though this may seem obvious—any evaluation method must be used correctly. Evaluation methods are misused a surprisingly high percentage of the time. Finally, data obtained from an evaluation must be correctly interpreted, which is a difficult task. Misinterpretation of evaluation results often leads to incorrect decisions.

The purpose of this book is to provide a reference for experienced or beginning evaluation practitioners, as well as for decision makers who must use the results of evaluations but are not themselves practitioners. This book emphasizes the concepts of mineral project evaluation rather than computational details. Various types of economic evaluation techniques commonly used—including conventional cost analysis, discounted cash flow, and option analysis—are described, as are their uses and relationships with geological, technologic, and financial evaluations. Commonly used decision techniques are also described. Additional definitions, formulas, and example calculations are given in the appendices.

This book discusses both the strengths and weaknesses of the various commonly employed evaluation methods. It is most important to recognize the many weaknesses in these methods. There is a distinct danger in believing too strongly in the results of project evaluations. Although the computing power is now available to ensure precise calculations, the resulting answers are often incorrect or misleading for a host of reasons. Nonetheless, the evaluation methods presented are the best available and are commonly used; when used correctly, they are powerful tools to help us understand the economics of investment projects.

This book is an outgrowth of lectures in mineral project evaluation and was originally designed to supplement texts used in my evaluation courses. Existing texts describe the details of discounting and cash flow analysis, but these books do not adequately describe the practical difficulties in conducting an analysis or interpreting the results or the use of alternative evaluation techniques. The meanings and uses of merit measures, such as net present value (NPV) or internal rate of return (IRR), as well as such analytical methods as probabilistic analysis, are not adequately described in many textbooks. In addition, the links between the determination of value and the decision-making process are not usually adequately discussed.

This book also differs from many existing texts in that it includes descriptions of the following:

- why NPV and IRR are both valid and useful merit measures
- the shortcomings of conventional NPV and IRR analysis
- the dangers of scenario analysis
- why conventional incremental analysis is not needed to correctly calculate IRR
- why the desire to achieve NPV-consistent IRR is misleading
- how probabilistic analysis is used in conjunction with risk preference theory
- how competitive cost analysis is conducted and used
- why option pricing is an important addition to the evaluation process
- how NPV-related merit measures suffer the same shortcomings as do IRR-related merit measures

This publication is intended to be a source book for decision makers and a supplemental text for courses in project evaluation. While specific subjects are discussed in detail and a broad range of topics are covered in this text, more detailed discussions of discounting and cash flow analysis can be found in numerous other engineering economy texts, such as Newnan (1988), Au and Au (1992), Steiner (1992), or Stermole and Stermole (1993). Evaluation of industrial and social projects are specifically addressed in a series of manuals developed by the United Nations Industrial Development Organization (UNIDO) (Gittinger 1982; Squire and van der Tak 1975; Ray 1984; Azzabi and Khane 1986; Behrens and Hawranek 1991; Dasgupta et al. 1992; Frohlich et al. 1994). Texts that address the financial evaluation of mineral projects include Gentry and O'Neil (1984), Boyd (1986), Newendorp (1975), and

Barnes (1980). Appraisal techniques used by assessors are described in Eckert (1990). Finally, decision techniques utilizing data obtained through the evaluation process are described in Keeney and Raiffa (1976).

Chapter 1 of this book gives a general overview of the evaluation process, including its goals, constraints, and uses. Aspects of measuring project worth and the role of the investor's perception of risk in project evaluation are given in Chapters 2 through 7. Chapter 8 describes the steps that often compose a project evaluation, as well as the potential errors that may be made, and presents conclusions and observations. Appendix A contains a compendium of simplified formulas, derivations, and examples that might be useful, and Appendices B and C contain descriptions of option-pricing techniques (written by Professor Graham Davis of Colorado School of Mines, Golden, Colo.) and certainty equivalence (written by Professor Michael Walls of Colorado School of Mines), respectively. There is also a glossary of commonly used terms in project evaluation.

Acknowledgments

I am most grateful for the comments and suggestions of those who have critically read this manuscript, for their input has been considerable. Not only did Mike Walls and Graham Davis each write an appendix, they contributed important improvements to the text as well. I particularly thank William van Rensburg (University of Texas at Austin), Dale Colyer (West Virginia University), Franklin Stermole (Colorado School of Mines), and Thomas Kaufmann (deceased; formerly of Colorado School of Mines) for their critical comments and helpful suggestions. Many thanks go to John Saymansky and Hong-Seok Jang, former students of mine, who critically read the manuscript and asked penetrating questions. I thank Mark Emerson, Leons Kovisars, and Ron Promboin for their comments in conjunction with offering short courses in project evaluation, as well as Mike Cramer, Jeff Kern, and Jerry Knight for their contributions to the descriptions of evaluation methods. Special thanks also go to Lisa Lewis and Gloria Nestor for preparing the text and illustrations. I also thank Phil Murray for his excellent editorial work in transforming my draft manuscript into a well-organized and easily understood book. Thanks also to my wife, Nancy, for her understanding and help in the demanding tasks of proofreading and providing moral support. Last, I also thank my students for their forbearance and their many questions, comments, and ideas. It is for them that this book was written. Any omissions and errors are fully my own responsibility.

CHAPTER 1

General Description of the Evaluation Process

An instruction book on project evaluation methods should accomplish two goals. The first is to present and describe the evaluation tools, such as discounted cash flow analysis, net present value, and rate of return. The second goal—one that is often omitted—is to describe how to use the tools to make investment decisions. Consider an analogy: A home builder needs to know how to hammer a nail and saw wood, but being familiar with the tools does not give him or her adequate knowledge to build a house. Both kinds of knowledge are required. This text attempts both to describe the required tools for project evaluation and to provide a blueprint for putting those tools into practice.

The evaluation process can be divided into two related phases: the valuation phase and the decision phase. The product of the valuation phase is a set of merit measures of project feasibility. These merit measures are used by decision makers to choose which project to undertake. The decisions are based on the principles of decision-making and operations research described in Keeney and Raiffa (1976). Since decision-making and operations research is a large field in its own right, this text focuses primarily on the valuation process and makes reference to the decision-making processes as needed.

To conduct a proper evaluation, analysts must first recognize the purposes of the evaluation, for whom the evaluation is to be done, the types of methods most suitable for specific needs, appropriate decision criteria, and what they can and cannot expect from the evaluation. They also need to recognize the strengths, weaknesses, and possible sources of confusion in the evaluation process so that they can correctly interpret the results. With so many variations in the types and goals of project evaluation, no single method is suitable for all cases. Consequently, there can be no predetermined, or "cookbook," approach to project evaluation.

A mineral project may consist of a developed property or a property that will or may be developed in the future. Consequently, at first glance, there appear to be two broad classes of mineral project properties for which evaluations may be necessary: (1) property that is conventionally considered to be an immediate investment opportunity and (2) property for which no immediate sale is contemplated. The perceived difference between these two classes is the reason investment analysis is sometimes considered to be separate from the appraisal of property's fair market value for taxation and other purposes. However, an investor always has the option to sell a property whether a sale is contemplated or not, and the value of any mineral property depends on the potential for that property to be developed, regardless of when such an investment may take place. Therefore, the processes of mineral project evaluation should be basically identical to those commonly used to appraise mineral properties for nontransactional purposes, such as determining value for taxation.

Thus, mineral project evaluations can instead be placed in one of the following two broad classes: (1) determining the fair market value of a property for taxation or similar purposes or (2) determining value for investment purposes. Fair market value can be defined as the value a willing buyer and willing seller may place on the property in the absence of any circumstances that would force the owner to sell or the buyer to purchase. The investment value can then be defined as the value at which a transaction would actually take place, in which case the actual investment amount may or may not equal the fair market value of a property. The difference between the two values depends on the specific investment requirements and constraints of the investor.

When an appraisal is desired to determine a property's fair market value for taxation, the purpose of conducting project evaluation is to provide data that the assessor can use to better determine value. When the evaluation is for investment decisions, the purpose is to provide data to help decision makers make better choices. In both cases the evaluation process simply provides data to individuals, who then interpret the data to determine the value of the property in question. However, given the imperfect nature of the evaluation procedure, which involves predicting future events, the information obtained will lead to better decisions only if the results are accurate. Therefore, to be useful, the results of a project evaluation must be more correct than incorrect, and the evaluator must be in a position to tell that this is so. Obtaining correct data and reaching proper conclusions are not easy. In addition, it is not easy to tell when data are incorrect, and it may be impossible to tell when conclusions are inappropriate. It is far worse to reach erroneous conclusions and believe them to be true than it is to have no information at all but realize this ignorance and act accordingly.

There are two types of errors with respect to undertaking a project: A project can be accepted when it should have been rejected, and it can be rejected when it should have been accepted. If a good project is rejected, all the benefits will be lost. Project evaluators usually place an emphasis on identifying the benefits. People are very motivated by the prospect of gaining from a project. However, much more serious problems can arise when a bad project is accepted—bankruptcy and ruin in the case of a private project or a disastrous outcome in the case of a social project. The effects of losing money already gained or money that a better project would have generated for the same investment are far more damaging than simply not acquiring future gains. Therefore, any project evaluation should pay at least as much attention to how much may be lost as to how much may be gained. Improperly analyzing the potential for losses is a serious deficiency of many evaluations.

The amount of attention an analyst should pay to the level of possible losses, given the possible benefits, depends on the investor's perception of acceptable risk. The level of acceptable risk varies among investors, is different for similar projects under different stages of development, and may be different for dissimilar projects. In addition, the level of risk may change for any project or investor over time.

Analysts may be tempted to conduct an evaluation process such that the results consist only of one simplistic measure that determines whether or not a project is worthwhile. The appeal of using a single merit measure is that a computer can select the most profitable project, thereby relieving the decision maker of this duty. Of course, this approach is improper. In most cases, analysts must consider many factors. Though computer-generated results are indeed valuable, it is up to the investor to interpret the significance of any results and to make decisions. The computer does not make decisions.

Even though a detailed project evaluation will not give absolutely correct results, it can still provide important benefits. The evaluation process forces the consideration of all alternatives, factors, and details, and it provides a framework to help organize thoughts. This alone justifies the time and cost involved in constructing an evaluation model. In addition, the evaluation process identifies the key factors that determine the success or failure of the project. Once analysts know the key factors, they can spend more time and money on trying to obtain better information on these factors. This enables the analysts to spend the evaluation budget more efficiently. For example, evaluators might be able to determine the costs of a project with an accuracy of •5.0%. If they can determine the future prices of the commodities to be produced with an accuracy of only •50%, they should focus their effort and money on improving price data or reducing price-related risks rather than improving cost data.

USERS OF PROJECT EVALUATION METHODS

There are three general types of users of project evaluation results: private investors (including privately owned corporations), lenders, and governments. Private investors are usually the project sponsors and operators. Lenders include commercial banks and institutional organizations, such as the World Bank, or downstream processors who lend money in exchange for exclusive rights to purchase the mine's products. Governments at both the national and local levels may be participants as lenders, contributors of equity, or taxing or regulatory agencies.

Each of these types of users has different investment and evaluation decision criteria. Therefore, each will interpret the results of a project evaluation differently, and each may use different evaluation methods. For example, private investors are generally willing to take risks and to receive compensation for their trouble. Lending institutions, by contrast, will accept less risk and will generally view a given project in a different light than a risk-taking investor. In addition, differences in regulations may cause banks and private investors to react differently.

Governments are concerned with a broader range of benefits and costs than are commercial banks or private investors, so they view project evaluation results and techniques differently. Governments must consider social as well as private welfare gains and losses. As a result, the diverse government objectives—economic development, employment and wages, government income and the provision of social services, the distribution of wealth, and environmental and social problems—are more likely to be in conflict than those of commercial banks or investors.

International lending agencies, such as the World Bank and state-controlled banks, have some of the characteristics of both lenders and governments in that they usually consider a broader range of benefits and costs. Not only must these lenders be conservative to protect their depositors, they must consider the social aspects of mineral projects as well.

Many mineral projects, and almost all large ones, involve all three of these types of users of project evaluation results. Not only must each party understand the project from its own perspective, each must understand the position of the others if an optimal, mutually acceptable position is to be reached. Therefore, each participant that has a financial interest in a mineral project must be aware of the evaluation methodologies and criteria of the other participants. For instance, if a private investor plans to approach a lender for funds, the evaluation must be prepared in a manner that is acceptable to the lender.

Despite the obvious need for specifying a clear set of objectives and defining criteria to measure the results of a project evaluation, these steps are often neglected in project evaluation. Corporations and governments may have many and often conflicting goals and objectives. In addition, one firm may have goals different from another's, even in the same industry. There are as many different sets of goals for a firm as there are numbers of firms. While profit maximization is usually the most important objective for a corporation, other possible goals include being the biggest, being the most innovative, or having the greatest market share.

Project evaluation by private investors has always been a keystone in the implementation of new mineral projects. Prior to the mid-1960s, most projects were financed by internal funds of the mining companies (Radetzki and Zorn 1980). Since then, capital requirements for very large mineral projects have become so high in relation to the capital assets of any single mining company that almost all these types of projects are now financed through debt and involve joint ventures. As a result, many financial institutions have become very heavily involved in mineral projects.

CHARACTERISTICS OF MINERAL PROJECT INVESTMENT DECISIONS

Before specific types of evaluation methods and decision criteria can be addressed, it is important to know more about the characteristics of the investment decisions at hand. These characteristics determine which evaluation technique ought to be used. It is possible to apply an incorrect technique or decision criterion and unknowingly obtain an answer that is mathematically consistent and may even seem reasonable but is still incorrect. It is also possible to do everything correctly and then misinterpret the results. Either of these situations may well result in incorrect decision making, which is the worst possible outcome of an evaluation effort.

The basic procedure of any evaluation is to compare the consequences or relative values of all possible alternative actions and then make informed decisions based upon the observed results. Determining the financial value of a capital investment involves comparing the expected returns from the investment with all alternative uses of the capital. The opportunity cost of capital—that is, the foregone benefits that would have been received from the next best investment—must be considered. For example, if the only choice is between keeping money in a bank accruing 6% annually and investing in a stock or a project, the investment must yield at least the 6% opportunity cost of capital to be worthwhile. The opportunity cost of capital is an important concept that is central to the evaluation methods described in this text.

The principles of evaluation explained in this text can be used equally well with nonmineral and mineral investments. However, mineral investments can have a number of characteristics that make them somewhat different from other types of investment opportunities, including the depletable and often unique nature of the ore reserves, the unique location and characteristics of the deposit, the existence of geologic uncertainties, the length of time required to place a mineral property into production, the usually long-lived nature of the operation itself, and the pronounced cyclical nature of mineral prices. In addition, mineral deposits can be mined only where the minerals are found. Therefore, options for locating a mineral operating site are likely to be more limited than with other types of industrial developments. This decrease in flexibility increases the risk of mining ventures compared to other types of investment opportunities.

Each mineral deposit is unique in terms of location and ore grade and characteristics. It theoretically cannot be replaced when depleted, although similar deposits can often be found or purchased. This aspect makes it difficult to compare the value of one mineral operation with another. Risk is increased since there is no guarantee that a search will yield a new mineral deposit to replace a depleted deposit.

The effects of time greatly influence the value of a mineral project, as they do any other long-lived investment. Usually prices and costs must be predicted, which introduces an element of risk common to most other types of investment opportunities. However, many mineral prices are cyclical in nature, and the difficulty in forecasting prices and costs poses particular problems in evaluating and planning mineral projects (Labys 1992).

Time also affects mineral projects in several ways that are not always present in other investment opportunities. First, significant mineral reserves are usually required for a long-lived mineral project. Since, by definition, reserves are quantities of ore to be mined in the future, the present value of a ton of reserves is less than the present value of a ton of ore that can be mined immediately. Many other investment decisions involve deciding whether to sell an item today or to wait until the future; thus, public officials who wish to tax mineral reserves to raise funds for public projects often misunderstand the fact that not all reserves have the same present value.

Second, the mineral evaluation process must consider that operating decisions made early during the life of a mineral deposit will affect the long-run value of the operation. For example, early mining of higher-grade ore increases early profits but lowers the average grade of the remaining ore and reduces the life of the mine. In addition, economic and operating conditions can unexpectedly change during the life of a mine, making flexibility of operation desirable.

Last, it is impossible to know the exact amount or grade of material to be mined until the deposit is depleted, for two reasons. The first has to do with geologic certainty. The geologic quantity and quality of the deposit must be determined by sampling, which by definition gives statistical estimates. The second reason involves economic certainty. The quantity and quality to be mined at any given time depend upon mining and processing costs and the prices of the resulting commodities. Since future prices cannot be forecast accurately, it is usually difficult to determine reserves even though there may be a high degree of confidence in the geologic-based estimates.

Project evaluation and finance are usually thought of in the context of opening new projects, but most decisions instead concern modifications of existing operations. Several basic types of decisions can be made concerning a mineral deposit. Each type of decision requires different sets of information and perhaps different interpretations of the evaluation results. Some typical basic investment decisions are (1) whether to do nothing or invest elsewhere, (2) whether to open a new operation, (3) whether to maintain, increase, or decrease output from an existing operation, (4) whether to close temporarily or permanently, and (5) whether to reopen.

These types of basic decisions are different in terms of identifying the relevant costs. For example, initial capital costs are relevant only for decisions to open or expand an operation. Maintenance, restarting, and cash operating costs are very important in deciding whether to close or reopen a facility. Exit costs and expectations about future costs and prices heavily influence decisions as to whether a facility should be permanently or temporarily closed.

These dynamic aspects of mineral evaluation cause many problems for project evaluators, engineers, and operators. Given the uncertain nature of many of the inputs, it is often difficult to determine, for planning purposes, such basic operating parameters as the optimal rate of extraction, average grade, and cutoff grade for a planned operation.

Investment costs and benefits for some projects are, at best, difficult to quantify and evaluate. For example, an investment may be necessary to preserve market share or an organization's position as a technological leader; the resulting benefits are intangible and thus hard to quantify. Investment decisions concerning human safety and pollution control are also difficult to evaluate unless failing to meet regulatory standards incurs a defined penalty, such as a fine or plant closure. Often, evaluations of these types of projects identify the least-cost options as the best options.

Many mineral investment decisions are forced upon organizations because of external events, such as an unexpected depletion of ore or changes in energy, labor, or government costs. In addition, investment decisions must often be made in a shorter period of time than desired, which severely limits the amount of information that can be obtained and analyzed.

Certainty Versus Uncertainty

A project in which future prices and costs are known with certainty can and should be evaluated in a different manner from one in which these factors are not known. For example, consider a coal-mining project in which the product is sold on long-term contract and mining costs are fairly predictable. Such a project can be evaluated by determining the identified value of future cash flows based on a discount rate that represents the investor's opportunity cost of capital—with no further increases in the discount rate to account for risk.

On the other hand, a gold mine with costs of $14.50/g ($450/oz t) would profitably operate only when the price of gold exceeds that amount. Since analysts cannot accurately forecast the price of gold, they do not know when the mine would operate or the amount of profit it would make when operating. In this case, determining the discounted value of unknown future cash flows gives only limited help to an investor. However, this does not mean that such a property has no value; rather, it means that the value must be determined in another manner.

When uncertainty and risk are relatively absent, project evaluation is a reasonably easy exercise. However, mineral projects often involve commodities for which prices or operating procedures are difficult to forecast, and analysts must make decisions concerning how best to evaluate such projects. The introduction of risk greatly complicates the evaluation process. The key to successful evaluation is the proper inclusion and quantification of risk and the consequences. There are numerous ways in which risk can be handled, as discussed in Chapter 5.

Mutually Exclusive Versus Independent Projects

There are two broad types of investment decisions based on the number and relationship of the projects involved. Often a decision maker must choose one project over another, such as deciding to build either a coal-fired or hydroelectric power generation plant. Since only one project can be accepted, these projects are defined as being mutually exclusive. On the other hand, investment opportunities that have no relation to each other, such as a copper mine and an oil field, are defined as being mutually independent. An investor can undertake either or both projects as long as each project meets minimum return requirements.

The ranking of projects is also affected by whether there are sufficient resources or capital to undertake all projects, even if the projects are independent. The usual case is that there are never sufficient resources or capital to undertake all projects that are available for investment. Thus, capital rationing is necessary. Under conditions of capital scarcity, the projects selected are those that yield in total the greatest benefits given the available budget.

Unequal Project and Service Lives

It is often necessary to compare the benefits and costs of two or more investment opportunities that have unequal lives. How to handle the unequal lives in an evaluation depends on whether project lives or service lives are involved. Those investments involving comparisons of alternatives that perform an identical service must consider the service life of the alternative investments. For example, it would be incorrect to directly compare the value of an investment in a machine that would last only 1 year with that for a machine that would cost more but would last 20 years. An obvious solution in this case is to compare the purchase of 20 inexpensive machines over time with the investment of purchasing the single more expensive machine. In practice, it may not always appear easy to account for service life differences for the purpose of comparing investments. However, if the analysis adheres to the theoretical evaluation requirements described later in this text, comparisons of projects with unequal service lives can easily be accomplished.

TYPES OF EVALUATIONS

There are many types of evaluation methods, partly because there are many needs for different types of evaluations and partly because of the inadequacies of many of the methods used. More than one method is almost always used to evaluate any single prospect.

Evaluation methods can be divided into two broad classes: (1) positive evaluation methods, which are based on measurable factors, such as economic efficiency, costs, and prices; and (2) normative evaluation methods, which are based on less easily measured factors, such as social values, ethics, and individual and collective value judgments. It is important to realize that these two distinct classes exist. It is easy to unknowingly mix the two and create much unnecessary confusion in the evaluation process. This is particularly true with government-evaluated social projects.[1]

1. The World Bank uses a slightly different classification scheme. In their scheme, a financial evaluation is one carried out from the perspective of a private investor; it includes both investment and financial analysis. On the other hand, an economic evaluation takes the perspective of society as a whole and considers all costs and benefits, including regional development benefits and external pollution costs. See Duvigneau and Prasad (1984).

Positive Evaluation Methods

Positive evaluation methods are based on positive economics, which is concerned with measurable factors based on economic efficiency, costs, and prices. Economic efficiency entails accomplishing a task at a minimum measurable cost or obtaining the greatest benefit from a given investment. A number of positive evaluation methods are described in the following subsections.

Geologic. Geologic quantities of material can be measured using geologic inferences and engineering calculations. However, ore and reserves are defined as material that can be removed at a profit given existing or immediately anticipated technologies and prices. This concept of ore gives rise to two common misunderstandings.

First, geologic reserves and ore grades are sometimes considered discrete and unchanging except for the effects of depletion. This misconception is especially true among public officials who have limited knowledge of mining and mineral values but are responsible for mineral- and energy-related fiscal and policy issues. However, the amount and grade of ore remaining depend on changing levels of geologic knowledge and varying economic conditions, as well as on depletion. Geologic reserves and cutoff grades of a mineral deposit are functions not only of geology but of extraction technology, costs, and prices. Changes in any of these variables will change the values of economical reserves and ore grade. This explains why companies may periodically report changes in the value of reserves and ore grade in excess of changes that would be caused by normal depletion and ongoing exploration.

The second misunderstanding has to do with the price associated with large tonnages of reserves. For example, it may be cited that, for a region that has 200 years of reserves of ore given the current extraction rate, if a ton of reserves to be mined immediately can be purchased at $1.10/tonne ($1.00/ton), then the value of the reserves is $1.00 times 200 years' worth of reserve tonnes. This is clearly incorrect; no investor could possibly afford to pay today the same per tonne value for material to be mined immediately as for material to be mined 200 years from now. Because of the time value of money, material to be mined so far in the future has little or no present value compared to material to be mined now.

The fact that the quantity of ore is affected by geologic knowledge, extraction economics, prices, and the time of mining has contributed to confusion on the part of public officials who attempt to appraise ore value for the purpose of taxation or zoning. The definition of ore also affects political decisions that are supposed to address mineral and energy depletion issues, especially in regard to so-called national security matters. The amount of material classified as ore at a given time and price by commercial firms can be considerably greater or less than the amount determined by public officials. Which quantity is correct depends on how one judges the details of the computations and assumptions. Differences are usually caused by varying assumptions concerning the effects of prices and production costs. Misunderstandings about the definition and characteristics of ore can easily lead to economically inefficient political actions.

In addition, geologic evaluations are always obtained through sampling and are, by definition, subject to sampling errors. Therefore, estimates of recoverable reserves, head grades, and ore and deposit characteristics that affect mining and milling costs and efficiencies are stochastic; they can be thought of in terms of probabilities and probable values rather than discrete and unchanging values. Better understanding of ore genesis and deposition, better drilling and sampling methods, and the use of computer programs in conjunction with geostatistics have improved the results of geological evaluation methods. For additional information on geologic evaluation and the determination of economic ore grades, see Peters (1978), Gentry and O'Neil (1984), and Lane (1988).

Technologic. Technologic evaluations can be divided into four main types: mining, processing, transporting, and marketing. The first three types, based on engineering estimates, are often the most sophisticated and most reliable of all evaluations. Engineering cost analysis can yield excellent forecasts of capital requirements and operating costs given forecast prices of the process inputs, rates of production, and correct geologic information. However, mining, processing, and transporting costs are dependent not only on engineering expertise but also on estimates of amounts that can be produced and sold, ore characteristics and mining conditions, and construction completion time. Failure to recognize all technological risks or relevant costs can easily lead to mistakes in forecasting. For example, governments may not recognize the extent of technologically related externalities (i.e., technologically related costs that are not included in the price of a good, such as the costs associated with preventing or allowing pollution) and exit costs (such as reclamation costs). From a social standpoint, these externalities should be identified, quantified, and included as costs in the evaluation.

Market and price evaluation is considered a type of technologic evaluation because it requires an understanding of the technical aspects of the uses and alternative sources of the product. This type of evaluation involves the prediction of future supply and demand for the product and is subject to a wide margin of error. It is particularly difficult to predict changes in supply technologies, which affect costs, and demand technologies, which affect quantity.

Investment. Investment evaluations use data from the geologic and technologic evaluations to develop cost, production, and price forecasts, as well as pro forma (i.e., forecast) cash flows. These cash flows are then used as the basis for determining valuation comparisons such as net present value (NPV), return on investment (ROI), internal rate of return (IRR), payback (or payout), comparative cost, and option value. These types of evaluations also identify possible risks, probabilities of failure and success, and probable profits and losses. Sources of error include the proper choice of discount rate; the proper choice of the combination of prices, costs, and quantities likely to be encountered; the effects of inflation; the timing of capital investments; and the possible effects of political and other risks and costs, such as taxes and exchange rates.

The specifics of investment analysis depend on the needs of the agent desiring the evaluation results. The three general types of agents generally involved in a project evaluation, as previously described, include private investors, governments, and lenders. Lenders include commercial banks, whose interests are fairly narrow; international lending institutions, such as the World Bank, whose interests include social as well as commercial concerns; and downstream processors, who want exclusive rights for purchasing the mine's products. Other types of lenders include export/import credit agencies, such as those that exist in the United States and Japan, who are interested in promoting international trade of their domestic producers and suppliers.

Since their needs vary, it is not surprising that these agents might use different methods and data in their investment evaluations. Private investors use actual market costs and prices to conduct investment analysis to determine financial returns on investment and financial risks. Governments must also include social costs and risks, which may be measured by shadow prices (i.e., prices that reflect true economic values in countries where government actions have distorted the market prices) rather than actual market prices. It is the government's duty to see that these costs are included in the project evaluation process. Commercial banks are interested in factors that affect the safety of the loans, while international lending institutions may combine their concern for the safety of repayment with social concerns of the host government.

Financial. The major task of financial analysis is to determine how to raise and repay funds needed to initiate and sustain the mineral project and necessary infrastructure. This is largely accomplished by assigning project risks to those participants who are in the best position to accept them. This arrangement results in the lowest overall level of risk and the lowest-cost financial package. For example, contract engineers take the responsibility for completion risk since they are in the best position to control project construction and completion problems. Other categories and responsibilities for risk may include production risk (producer), engineering risk (equipment providers), political risk (insurance company), market risk (producer and lender), reserve risk (producer and lender), environmental risk (producer and lender), and hazard risk (insurance company). For a discussion of identifying and assigning financial risks, see Tinsley (1985a, 1985b).

Funds for a project may come from a variety of sources, including retained earnings of the participants, sale of common and preferred stock, commercial and institutional loans, sale of bonds or other debt instruments, futures sales of commodities to be mined, and various types of loans arranged around purchases of equipment and services. A major portion of the financial analysis involves structuring the deal in a way to guarantee that all loans will be repaid. Since financial analysis builds upon the results of the geologic, technologic, and investment analyses—with all their errors—the chances for errors to occur at this stage are compounded. This creates a dilemma because lenders are highly risk averse and operate on very small margins, while mineral project evaluations at this stage are fraught with uncertainties and almost certainly contain unrecognized and compounded errors.

Normative Evaluation Methods

Normative evaluation methods address issues of equity, welfare, ethics, and value judgments. Examples of evaluations requiring these types of methods are environmental assessments, assessments of the effects of wealth redistribution, and determinations of the value of training and infrastructure to change the social well-being of a population. These issues are of the highest importance to the project's host county, but they cannot be as easily quantified or measured as market-based prices and costs. The social effects of a project must be evaluated in light of the social goals of a country. This is a particularly difficult task since many goals might be ill defined or conflicting and many of the social effects of a project are difficult to measure. However, no matter how difficult this evaluation is, the long-run success of a project depends upon a favorable social evaluation. For a more detailed explanation of normative evaluation methods, see Dasgupta et al. (1992).

END PRODUCTS OF PROJECT EVALUATION

Saying that the purpose of a project evaluation is just to determine the value of a project is an overly simplistic description. Project evaluation yields a number of end products that decision makers use to determine project value and make investment decisions. The most important of these end products are briefly described in the following paragraphs.

Identification of All Possible Alternatives

It is incorrect to evaluate a single project as though it were isolated from the rest of the economy or the firm. There are always alternatives to any decision. To correctly determine the value of any single project, the values of all alternatives must also be determined. For example, a viable alternative may be to do nothing or to wait until more information is obtained. Simply identifying the possible alternatives is always a first step in the evaluation process and provides much information for the decision maker.

Pro Forma Cash Flow and the Identification of All Factors

A pro forma cash flow model provides a forecast of possible future investment needs and cash surpluses under different assumptions concerning prices, costs, and quantities. It provides a basis for additional testing and understanding of the project. Constructing this model helps an analyst to organize thoughts and to make sure that all relevant variables and factors are included in the analysis.

The inclusion of all relevant costs and benefits in the evaluation is of particular importance in the construction of a cash flow. Although the cash flow format itself helps analysts to identify and include all benefits and costs, omissions can still easily be made. For example, analysts may not account for shutdown and abandonment costs in many instances. In addition, the benefits from a new activity may represent a cost to another ongoing activity, and both these costs and benefits need to be recognized if a true net value is to be determined. Such an instance might arise when mining technology is changed to increase productivity but also results in the need for greater processing and waste disposal capacity. Pro forma cash flow models are described in more detail later in this book.

Identification and Quantification of Risks

A complete understanding of the risks, their consequences, and their probabilities of occurrence is an absolute necessity for any project evaluation. A pro forma cash flow can provide the basis for risk assessment in an investment analysis, as well as a means to determine the optimal assignment of risk in a financial analysis. The goal of risk assessment is not to reduce risk–that cannot be done by analysis–but rather to increase the understanding of risk so that appropriate action can be taken.

There are five components to risk assessment:

1. To identify the risks.
2. To determine the consequences of the risks taken.
3. To determine the likelihood of the events actually happening.
4. To determine the investor's attitude in taking the risks so that the risks may be properly evaluated.
5. To actually evaluate the risks.

The need for the first component requires no explanation. The second and third components are needed because evaluators must be able to differentiate between (1) risks that have high probabilities of occurrence but cause small changes in project value and (2) risks that have low probabilities of occurrence but cause drastic changes in project value. As an example, paying a low price for an exploration right for a property that has a low potential to contain a mineral deposit must be differentiated from paying a potentially ruinously high price for the right to a property that has high potential. Failure in the first instance would not have a great financial effect on the investor, whereas failure in the second instance would cause the investor to go into bankruptcy. The fourth component, the attitude of the investor toward risk, is needed to enable the evaluator to properly assess the value of a risky project. This type of analysis involves the controversial concept of preference theory, which is discussed in Chapter 5.

The determination of risk is becoming increasingly complex in the minerals and energy industries because of the need to include environmental risks and costs in project evaluation. Good measures of environmental costs or benefits are often not available, and it is difficult to include these issues in a conventional pro forma cash flow analysis. However, all costs and benefits and their associated risks, including environmental factors, must be included if the evaluation is to be complete. For additional readings on risk, see Glickman and Gough (1990).

Risk and Return

In general, the higher the risk experienced by an investor, the higher the expected returns. Without the promise of higher returns, an investor would have no reason to consider projects with higher risks. If risk minimization were the investor's goal or if there were no premium paid for risk assumption, the investor would choose only low-risk securities, such as U.S. Treasury bills. If governments in countries perceived as high risks by investors ignore or neglect this aspect of investment theory, they will see little new investment.

Although risk in the mining industry has changed over the past 20 years, it is still present and represents an important consideration in the evaluation of any project. As R.P. Wilson, chief executive officer of the RTZ Corporation observed, mining is more exposed to political and economic shocks than many other industries, but rewards go to the risk takers ("A Closer Look" 1992).

The concept of rewards going to risk takers is easy to understand. However, the critical problems involving measuring the risks, rewards, and penalties associated with a particular project and identifying the investor's attitude toward risk are generally complicated. A number of methods used to measure the various aspects of project risks and investor acceptance of risk are described later in this book. For further information on risk, see Newendorp (1975) and Stermole and Stermole (1993).

Identification of Critical Variables

Having a model allows evaluators to conduct sensitivity analysis to identify which variables are most critical to determining the value of a project. Evaluators can then focus attention on these variables to make sure the projected values are as accurate as possible. They can also judge the probability of deviations from forecast values and test the consequences of these changes.

Identifying the critical values sometimes leads to surprises. For example, it is not uncommon to find that analysts can predict capital and operating costs with a fair degree of accuracy, but they usually know much less about future product prices. Sensitivity analyses of mineral projects frequently show that a given percentage deviation in projected capital or operating costs has about half as much effect on the value of a project as the same percentage deviation in projected product prices. Analysts can be surprised to discover that, despite great expenditures for a thorough technical analysis, more information is often available on less critical variables, whereas there is little information on the most important variables, which are likely to be product prices.

On close examination, some variables, because of their short-lived impact, turn out not to be as critical as they first appear. For example, unexpected increases in operating costs may persist throughout the entire life of the project, whereas prices are expected to fluctuate. Therefore, unexpected deviations in yearly prices may not be as critical as a persistent unexpected deviation in operating costs.

Identification and Quantification of Value or Rents for Taxation

One of the major goals of a host government in promoting a mineral project is to obtain maximum benefits or tax revenues from the project. Certain difficult questions always arise: How should the government participate in the project? How much should the project be taxed? How many investment incentives should be given? The key to providing answers is in identifying the amount of economic rent in both the short and long runs. Economic rent is defined as that amount of profit that can be taxed away without affecting decisions of the project investor. It is based on the difference between received price and total production costs. The key is how total production costs are defined. Economists include in their definition of cost a provision for the recoupment of all capital invested plus a minimum rate of return for the investor's trouble. There is also a distinction between short-term rents (or quasi-rents) and long-term rents. Short-term rents can be taxed away without affecting the immediate operations of an existing project. However, extracting these short-term rents will have long-term effects on future investment. Governments considering whether to tax short-term rents should be careful.

Cash flow analysis and cost models can help provide answers to the aforementioned questions. Some rent capture methods are more efficient than others in extracting rents from mineral projects. Tax and resource economics theory provide direction in choosing taxation methods. For additional information on taxation and rent capture, see Blackstone (1980), Fletcher (1985), Foley (1982), Garnaut and Ross (1983), Gillis (1982), Schenck (1984), Torries (1988), and Broadway and Flatters (1993).

Financial Optimization

Once a decision has been made to undertake a new project or to invest in an existing project, the necessary funds must be made available. Although there are only two basic sources of investment funds, equity and debt, there are many variations, constraints, and conditions that complicate the process of choosing the optimal combination of types and amounts of debt and equity. The point of financial optimization is to identify and obtain the combination that results in the lowest cost of investment funds for the project.

Geologic, technologic, and investment analyses are all used in conjunction with finance theory to determine the optimal financial structure of a project. If debt is a desired way to raise capital, lenders may conduct their own independent geologic, technologic, and investment evaluations before agreeing to provide funds.

CAPITAL BUDGETING AND SCARCE RESOURCES

Capital budgeting involves allocating scarce (i.e., limited) resources by project participants (Bierman and Smidt 1966). All project participants have scarce resources, such as capital, foreign exchange, managerial and supervisory expertise, natural resources, amount of infrastructure, and trained labor. Evaluators must rank projects to identify those that are the most desirable and affordable given existing budgets and goals. Trade-offs then become inevitable since a choice must be made from among many projects with their assorted benefits and costs. Trade-offs must commonly be made among such factors as low initial capital requirements, early positive cash flow, high taxes, low project risk, high foreign exchange earnings, rapid sales growth, low production costs, low environmental impact, high labor utilization, encouragement of vertical or horizontal expansion, and high return on investment. Obviously, investors cannot have the best of all these criteria at the same time. The problem is in determining the acceptable trade-offs.

DYNAMIC NATURE OF PROJECT EVALUATION

Project evaluation is a continuing process rather than a one-time event. Numerous projects compete for the same scarce resources at any given time. Changes in the budget, evaluation criteria, or costs or benefits of any of the competing projects may change the evaluation results and ranking for any single project under consideration.

Steps in Project Evaluation

The actual steps in evaluating a project vary somewhat depending on specific circumstances and the size of the investment. In general, the larger the investment, the more thorough and lengthy the investigation is likely to be. For example, the purchase of a small mining property adjacent to an existing mine may require only a few weeks, whereas the purchase of a larger set of properties may take as long as several years.

While there is a theoretical evaluation path that takes into account all relevant variables, the actual path may vary in practice because of a lack of time, a lack of information, or the influence of a project mentor or decision maker. This section will first discuss a theoretical path that accounts for all factors and will then turn to describing realistic alternative paths.

As is shown in Figure 1.1, the first step in project evaluation is to define long- and short-term interests and goals. It may seem unnecessary to stress the importance of setting realistic goals, but many projects have failed because clear goals and evaluation criteria were not adequately developed prior to the project evaluation process. Many projects have been undertaken only to be subsequently abandoned or sold at a loss because they were not consistent with the overall goals and acceptable risks of the investing company.

To be attainable, goals must be consistent with the resources available. Therefore, any actions must take place within the constraints imposed by the financial budget, availability of managerial expertise and skilled and unskilled labor, and the political environment. These constraints change over time and depend on the needs and performance of ongoing projects, as well as the needs of potential new projects. Sources of change in the constraints are linked to the constraints by dotted lines in Figure 1.1.

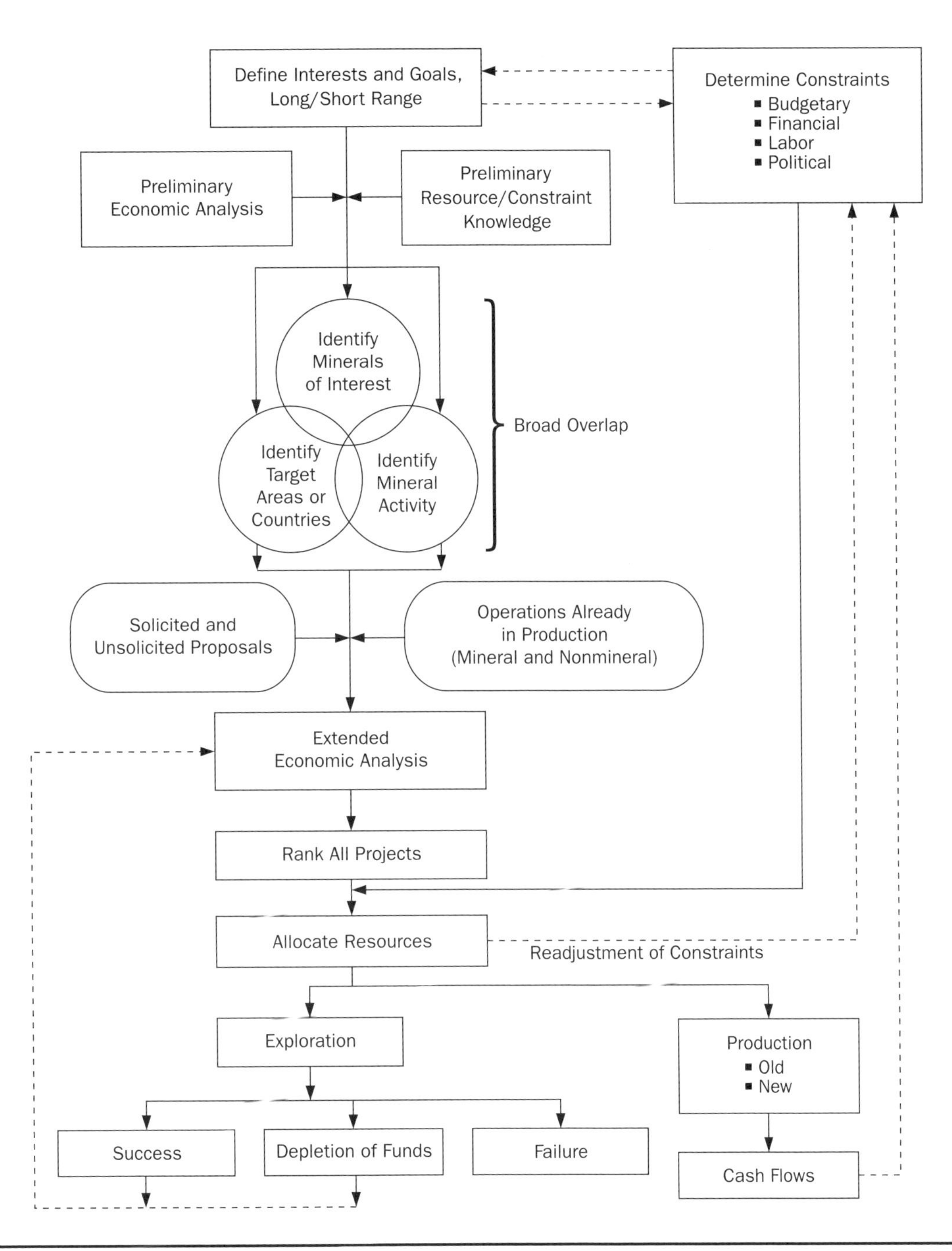

FIGURE 1.1 Theoretical dynamics of mineral project evaluation and implementation

A new project must first be evaluated on the basis of preliminary information in light of the known project parameters, existing goals, and constraints. The corporation or government must then decide how best to participate in mineral development. For example, a private firm may decide to seek only certain minerals in selected countries. Numerous firms followed this path during the 1980s when they decided to seek gold only in North America and Australia.

A firm or government must also decide how to be involved in the mineral industry. There are many options, such as stressing exploration (and selling developed prospects), producing (which includes contract mining and processing), or marketing (which includes mineral brokering). Each type or combination of activities carries a different set of risks, costs, and benefits for the operation given the existing goals and constraints.

As previously discussed, there are more projects that could be undertaken than available resources will accommodate. Sources of these projects include both solicited and unsolicited proposals, as well as proposals from existing operations. Quite often, incremental investments in existing mines or plants yield very high returns. In addition, ongoing investment is almost always required simply to sustain an existing operation. Firms and governments may also consider investment in fields not related to minerals. Investment in agriculture or transportation facilities may yield even higher returns than investment in a good mineral venture.

Only after all possible projects and options are identified can the results of extended economic analyses of the individual projects be compared. Given the goals, constraints, and characteristics of each of the prospective projects, the projects can be ranked. Then, given the budgetary constraints, decisions can be made and resources can be allocated to the most favorable project or groups of projects. Allocating resources among new and existing projects then changes the original constraints, as shown in Figure 1.1.

The decision to put capital into exploration can lead to one of only three outcomes: failure and abandonment of the prospect, depletion of available funds, and success. Obviously, only success may lead to near-term benefits. Both depletion of funds and exploration success require another round of decisions about the allocation of funds given the new set of constraints.

The evaluation of any single project does not take place in a vacuum. It depends on the value of all other projects under consideration and the relevant constraints. The number and types of projects being considered, as well as the goals, constraints, and knowledge, all continually change over time. Therefore, project evaluation is an ongoing and dynamic process.

The theoretical process just described may have other configurations in real-world practice. It is possible that many of the comparisons and feedback loops will be bypassed because of a lack of time or the influence of a project mentor or decision maker. Conversely, it is not altogether uncommon for a decision to be indefinitely postponed as more and more data are gathered in response to rapidly changing market or economic conditions. This need for more information may lead to justifiable delays or project postponement. However, since perfect information is unattainable, the continued quest for additional information may lead to a result sometimes described as "paralysis by analysis." While these cases of accelerated or decelerated development are quite different from each other and the theoretical model, they are not necessarily better or worse. The strong influence of a mentor or decision maker may be justified in some cases, especially when that person has the experience to recognize that particular uncertainties may defeat a promising project. Similarly, paralysis by analysis may not be an undesirable outcome during a period of changing economic or financial conditions. However, given the rapid rate at which economic and market conditions change in today's business world, prompt decision making is often preferred.

The Value of Information

Obviously, the more information available on which to base a decision, the better the decision should be. However, since investors can never obtain all the information they would desire, they must make decisions on imperfect information. The problem then is whether to make the decision on the basis of information available or to gather more information. Since information is costly and delays in implementing a project reduce potential profits, this issue is not trivial.

The economist's response is that data should be gathered until the cost of the last bit of data collected just equals the benefit derived from that data. While this is theoretically correct, determining the marginal costs and benefits of data collection is extremely difficult in practice. For example, if a project were to be turned down on the basis of excessive geologic risk, more exploration could possibly reduce the risk to an acceptable level. However, while evaluators can forecast the cost of additional exploration, they cannot be certain of the value of the additional exploration results.

There are several approaches to determining the value of additional information. According to Holloway (1979), information has value only if it influences the actions of the decision maker. Therefore, if a project is driven by mentor or decision maker, the value of additional information may be low. If the availability of new information does result in a change in action, the value of information depends on the amount of uncertainty, the profit, and the risk attitude of the decision maker. In general, if the decision maker is risk neutral, the expected value of information is the difference between the expected value of the best alternative with information and the expected value of the best alternative without information. If the investor is not risk neutral, the value of information is equal to the cost that makes the best alternative with information equal to the best alternative without the information, which is best determined by trial and error.

According to Rice (1974), the key variables should be determined using sensitivity analysis (which is discussed in more detail later in this book). Following this, the key variables should be tested to determine how precisely they need to be known. The task is then to determine the probable loss due to insufficient knowledge concerning the key variables.

Another approach to determining the value of information is to determine the maximum value of having new information, even though analysts may not know the exact cost to obtain this information. This procedure sets an upper limit to what can be spent for additional information.

For additional discussion on the basis of decision making and the value of information, see Howard (1968), Rice (1974), and Holloway (1979).

CHAPTER 2 Nondiscounting Methods

There are six nondiscounting methods to determine the worth or feasibility of an investment opportunity:

- comparing the value of similar operations
- assessing the breakup value of an existing operation
- determining the replacement value
- comparing the earnings of existing operations
- calculating payback from an investment
- determining costs

These methods are discussed in the following paragraphs.

COMPARABLE (FAIR MARKET) VALUE

While there are many ways to determine the value of a project, one of the most reliable—and the one most likely to be accepted to resolve legal disputes—is based on the price as determined by actual market transactions. The true value, or fair market value, of a project is defined as the price willing and knowledgeable buyers and sellers in a competitive market are willing to pay.

Variations

There are at least three important variations of this method. The first is a broad-based approach that consists of determining the value of an entire project by comparing it with the values of similar projects under similar circumstances. Houses or equipment are often valuated in this manner. The value of a house can be determined by the recent sales prices of similar houses in similar neighborhoods. This method of determining comparable value is more difficult when applied to mining projects, however, because many projects have a number of unique characteristics that complicate direct comparisons, such as quality and quantity of ore, mining and processing costs, production quantities and products, location, and time of mining.

The second variation is very similar to the first except that value is determined on a per unit basis, such as value per tonne or barrel. For example, a house value might be expressed in terms of dollars per square meter or square foot of living area. A low-sulfur coal property might be evaluated by using historical per tonne prices of sales of low-sulfur coal deposits in other similar locations. Differences in the mineral and property characteristics are reflected in the unit value of the mineral. This method has been used to determine the value of large coal and oil projects and companies, as well as certain precious metal deposits and companies. It is commonly referred to as the royalty determination method of property worth or the comparable transactions method.

The time of extraction is also explicitly included in this variation. For example, as of 1997, similar coal properties that are in a developed coal-producing region, such as West Virginia, and are to be mined immediately in the future command prices or royalties of around $1.10 per tonne ($1.00 per ton) in the ground. Coal properties to be mined starting approximately 20 years later command royalties on the order of $0.17 to $0.22 per tonne ($0.15 to $0.20 per ton). The difference in value is due entirely to the time value of money and the greater risks involved in purchasing reserves to be mined at times farther in the future.

It is worth noting that even if the mineral properties in a specific region were identical in every way, the time of mining may be different simply because it would not be possible to mine all ore in the region instantaneously. In this example, as in all cases, it is improper to use the comparable value method without taking into account the time of extraction. This is a common mistake governments make in determining the value of reserves for taxation purposes. For example, West Virginia has huge quantities of coal and coal is a major source of taxes in the state. There is a popular conception among many citizens that since the state has 31,820 million tonnes (35,000 million tons) of minable coal and since coal to be mined in the immediate future sells for about $1.10 per tonne ($1.00 per ton) in the ground, the unmined coal in the state has a total taxable value of $35,000 million. However, it would take over 200 years to mine all the coal in the state at the current rate of mining. Consequently, the price investors pay now for coal varies from $1.10 per tonne for coal to be mined immediately to progressively less for coal to be mined in the future. Coal to be mined 10 years from now commonly sells for $0.11 per tonne ($0.10 per ton), and coal to be mined 20 years from now sells for about $0.011 per tonne ($0.01 per ton). Coal to be mined 40 or more years from now sells for even less. As a result, the total value of unmined coal in the state is about $2,000 million rather than $35,000 million.

The third variation of the comparable value method is to compare the value of two or more publicly traded companies on the basis of stock prices. If one of the companies is not publicly traded, financial and performance ratios taken as indicators of stock worth can be determined and compared. Market valuation depends not only on what the market thinks of the project or company in question but also on what it thinks of the project or company's value relative to all others in the market. This method of valuation, though it does have application, may not give a true indication of the value of an undeveloped mineral project because of a lack of information or an overly optimistic assessment of the prospect.

Important Considerations

It is essential to correctly interpret the intent of the buyer or seller in transactions assessed by the comparable value method. A comparable sales analysis is usually applicable only if the transaction value represents a fair market value of the property in question. Fair market value can be defined as the value a willing buyer and willing seller may place on the property in the absence of any circumstances that would force the owner to sell or the buyer to purchase. An actual transaction value may differ from the fair market value if the buyer or seller has special interests or constraints. For example, conditions involving forced liquidations or condemnations may cause the seller to accept less than the fair market value of other similar properties. On the other hand, when parcels need to be purchased to satisfy unique needs—for example, to acquire the last parcel for completing a longwall mining panel of an active mine—the buyer may have to pay more than the fair market value.

Another major problem with the comparable value technique involves identifying just what was transferred in the transaction and making sure that the types of assets transferred correspond with those of the property to be evaluated. For example, it is not unusual for a property containing both minerals and surface improvements to be purchased for the value of the minerals alone. For example, a coal stripper might buy coal with no thought of the value of the surface improvements since these improvements would be destroyed in the mining process. In this case, the value of the improvements should not be subtracted from the sale price to determine the coal value. In practice, both the buyer and seller would need to be contacted to allow the exact nature of the transaction to be judged correctly.

When an outright purchase of all assets (a fee simple transaction) occurs, the minerals and the right to mine the minerals are usually both transferred. On the other hand, the minerals alone may be purchased if the mineral rights have been severed from the land. In such cases, it may be necessary to determine the value of the right to mine the minerals as well as the value of the minerals themselves in order to conduct a meaningful comparable value analysis. This is often the case in the strip-mining of coal; the coal and the surface commonly have different owners, and the right to mine the coal often equals the royalty value of the coal itself.

The values of other assets—such as valuable timber, a property location desirable for future high-value use, and valuable buildings that would not be destroyed during the mining process—must be determined and excluded from the property transaction price to determine the value of the mineral. This is an imprecise process in which errors are to be expected. The only way this problem can be fully mitigated is for there to be a sufficiently large number of properties that a statistically significant average value can be determined. In this manner, mineral values determined for any given property may be in error, but the average value will be correct within a known range of values.

For an example of the use of the comparable market value method to determine the worth of a coal property, see Lipscomb (1986); and for determining the worth of a mineral property in general, see Boyd (1986).

BREAKUP VALUE

The minimum value of an existing project or company can be estimated by determining the total value of the individual disposable assets. This value is known as the breakup (or book asset) value and can be assessed in three ways: (1) determine the instantaneous liquidation (scrap) value if the assets were to be sold quickly in an emergency situation; (2) determine the auction value of the assets if they were to be sold in an organized fashion; or (3) determine the value if the assets were to be sold separately. The value of the assets disposed in these three methods can be estimated by an experienced appraiser giving an actual market assessment or, less accurately, based on the book value.

A variation of this method is to rank projects on the basis of Tobin's q, which is the ratio of asset value to stock price of a company. The higher the value of q, the better the acquisition. However, this method cannot generally be used to determine the value of undeveloped mineral deposits. For additional readings in the use of Tobin's q, see Tobin and Brainard (1977), Tobin (1978), and Lindenberg and Ross (1981).

REPLACEMENT VALUE

Replacement value has been widely used to value assets but sometimes with disastrous results. The rationale behind this approach is that purchasing an asset at less than its replacement value (i.e., the amount it would cost to replace the asset with a comparable asset) is always desirable. However, this is not always an accurate indication of value, especially when the location or technology of the asset is fixed, such as with a mine and beneficiation plant. A poor mine and plant with obsolete technology is no bargain even at a cost below its replacement value. Further, markets may change, and to replace a reserve or plant at the value determined by prior markets may not always be correct.

Nonetheless, reserves can be valued on the basis of replacement value. For example, the cost of purchasing reserves can be compared to the cost of exploration to justify either purchasing the reserves or conducting exploration activities. These uses of the replacement value concept can be valid if proper care is taken in comparing reserve quality, location, quantity, and risks.

EARNINGS MULTIPLES

The breakup value of a project places a value on the independent components of the project, but it ignores the future earnings capabilities of the assets in place. One commonly used way to determine the value of future earnings of a corporation or project is to value an asset at some multiple of its recent yearly average earnings. The multiple used reflects the risks and future potential earnings of the firm or asset. Higher multiples mean the venture is perceived as having higher risks but greater gains if successful. This is why growth stocks are commonly traded at prices in the range of 20 times earnings or greater, compared to average stocks, which trade in the range of 7 to 12 times earnings.

PAYBACK

The number of years required for the cumulative cash flow of a venture to reach zero is called the payback (or payout) period. This has been a commonly used evaluation tool for many years. The rationale of this method is that a shorter time required to get back the investment is better. While this method recognizes in an indirect manner the time value of money, it does not take into account the full stream of benefits over the life of the project. Payback is, however, commonly used as an effective means of risk assessment. In many cases, the farther into the future a forecast must reach, the more likely the forecast is to be wrong. Therefore, one way to reduce risk is to get the invested funds back as soon as possible, thereby preventing losses caused by unpredictable future events. In general, projects with payback periods in excess of 7 to 8 years are considered to be risky or to give low returns on investment.

Since this method does not take into account the total cash flow or distribution of cash flows over the life of the project or the time value of money, it is an inappropriate evaluation technique if used alone. It is better used as a screening criterion or constraint in conjunction with another evaluation technique, such as net present value.

For example, consider the use of payback in assessing the feasibility of developing a rich deposit in a remote and politically unstable area. The project may have a very attractive rate of return, but management will probably not give approval until it is shown that payback can be achieved in less than 2 years. Management is usually comfortable in investing under such conditions only if the investment could be recouped in a very short period of time.

COST ANALYSIS

Low operating cost is one of the keys to a profitable mineral project. If a project's average production costs are less than those of competing operations, the lower-cost operation will probably survive periods of low prices and low demand better. Therefore, if an operation's costs are in the lower portion of the industry cost curve, evaluators might feel safe about the project's chances of survival, even though future factors such as market and price changes cannot be forecast. In addition, once the industry cost curve is identified, an estimate of demand gives an estimate of market prices for the commodity. Knowledge about the industry cost structure is a powerful tool in project evaluation.

The key to conducting a successful industry cost analysis is to identify and treat the various cost items of each producer in the industry in a consistent fashion. This task is more difficult than it may at first seem because (1) corporations in different countries have different accounting and taxation methods and (2) costs are commonly accumulated in different manners at different stages of production within a single operation. Cost categories and terms commonly used in cost analysis are illustrated in Figure 2.1.

It is important to correctly account for all costs to derive a meaningful cost analysis. However, this task is not always accomplished in practice. For example, the costs associated with political risks, unexpected environmental or social factors, and abandoning an operation are meaningful but are frequently omitted. Note that these costs *are* included in the cost categories of Figure 2.1 and are not, in themselves, line items.

Cost Item	Description
+ Labor + Energy + Materials and Supplies + Local Taxes and Insurance + Transportation + Royalties	These are the direct costs required to produce the mineral product and deliver it to the consumer. All costs must reflect appropriate exchange rates.
Direct Operating Costs	This indicates the efficiency of the operation in terms of moving and processing rock. The effects of ore grade are suppressed.
– By-Product Credits	The actual prices received for by-products sold.
Net Production Costs	This indicates the effects of operational efficiencies as well as ore grade.
+ Sales, Overhead + Interest + Other Cash Costs	These are the cash costs that the operation can control. For example, interest can be swapped for debt or even be forgiven in cases of extreme financial hardship.
Cash Break-Even Cost	This represents the cash costs necessary to sustain operations. It is the cost basis on which decisions are made to shut down or reopen in response to changing prices and profits. It is the relevant cost in determining the financial viability of an existing project.
+ Initial Capital Recoupment + Return on Investment	Capital costs include recoupment of invested capital plus a minimum return on the investment. For decisions regarding future actions, these costs are irrelevant for existing operations but must be included for new operations.
Total Cost	This is the relevant cost for determining the financial viability of a new project.

FIGURE 2.1 Cost categories

Uses of Cost Analysis

Cost analysis is used as a basis for predicting the actions of the individual producers that make up an industry. Since past capital expenditures are sunk costs, they do not influence the future actions of existing producers. Therefore, the appropriate costs to determine future actions of either existing or new producers include only cash operating costs and a provision for future necessary capital costs. To correctly compare the total costs of existing producers with those of a new producer, past capital investments must be omitted from the costs of the existing producer.

In some cases, a more appropriate cost comparison would include an analysis of the cost of using the product. For example, in the case of steam coal production costs, given the differences in coal quality and the resulting costs of generating power, the relevant cost for judging competing coals is the total cost of generating electric power.

Cost analysis can also be used to determine the amount of economic rent that may be present in any single operation. Proper identification and quantification of economic rent are of paramount importance to governments because they form the basis for determining the level of taxes that can be extracted without affecting the producer's investment decisions. Economic rent for any existing operation is defined as the difference between the cash break-even cost and the price. Economic rent for any new operation equals price less total cost, which includes recoupment of capital plus some minimum rate of return on investment. For more details on the use of cost analysis and identification of economic rent, see Torries (1988).

When an industry is in a situation of oversupply, which is often the case, price is strongly influenced by the cash break-even cost of the highest-cost producer whose product is necessary to fulfill demand. Under these conditions, a cumulative industry cost curve represents the industry supply curve, and the intersection of the cash break-even cost curve and the demand curve identifies a reasonable and attainable market price of the commodity produced (although other factors, such as market speculation and size of stocks, also affect commodity prices). An example of a set of cumulative cost curves for nickel and forecast price is shown in Figure 2.2. Price is determined by the intersection of the cumulative industry cash break-even cost curve and the demand curve. Note that the cumulative cost curve differentiates among nickel producers who are and are not influenced by the price of nickel. South African platinum producers and Russian nickel producers (prior to 1994) produced nickel regardless of the nickel price.

Problems and Misconceptions

Cost analysis is a powerful evaluation tool, but it is not without problems and misconceptions. First, as pointed out by Adams (1991), all operations aspire to be in the lower cost category, which is obviously not possible. Thus, not all new projects can use low cost as the investment decision criterion. On the other hand, high-cost operations can be profitable if prices are increasing.

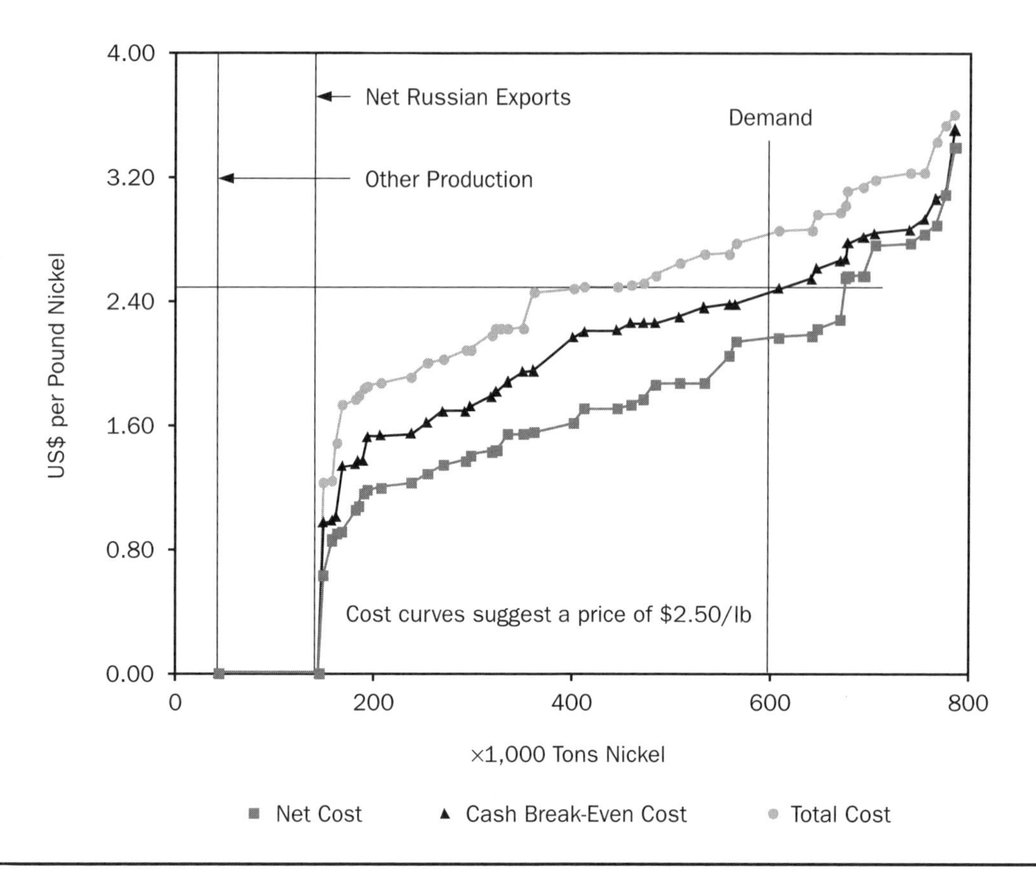

FIGURE 2.2 Example of cumulative cost curves and price determination

Cost ranges for all operators are more similar in times of depressed markets than during expanding markets, so the importance of low cost may change with market conditions and competitive positions of the producers.

Second, some cost analyses may not recognize that operations and costs are controllable, sometimes to a large degree, and that costs and production can be adjusted to meet the requirements dictated by the market and prices. In times of low prices, managers may be able to shut down high-cost and low-grade portions of their operations to maintain the necessary profit margins. When prices increase, the higher-cost and lower-grade operations can be reopened. However, cost analysis is an even more powerful evaluation tool when combined with the realization that operators adjust to changing economic conditions.

Third, cost analysis does not relieve the decision maker of having to make difficult forecasts of future prices. It requires that forecasts be made of prices of input factors, such as labor and energy; these forecasts, like those of product prices, are subject to uncertainty. However, the advantage is that fewer prices need to be forecast, and perhaps the prices of inputs can be forecast with a greater degree of accuracy than those of the mineral products.

Another concern is that—because of accounting methods, short-term exchange rate changes, or the presence of government subsidies—actual economic costs may not be apparent to the producer. In these cases, the apparent cost, not the true economic cost free of any distortions, dictates producer actions. However, the true economic cost is important to governments and to international lending and development agencies to ensure maximum economic development and efficient allocation of resources within the country.

Another problem is how to treat the benefits and costs of joint products. This problem is solved by determining appropriate by-product credits that can be subtracted from direct operating costs to determine net direct operating cost for each operation. This sometimes raises the question of what is a by-product and what is the main product, as with the case of a lead–zinc operation where both are produced in equal quantities.

Software Packages

Cost analysis is commonly used at the prefeasibility and feasibility stages to see if the probable capital and operating costs of a mine or project are within the realm of economic possibility. A number of commercial cost estimation packages are available that convert geologic data into reserve estimates and capital and operating cost estimates. PREVAL is a software package developed by the U.S. Bureau of Mines to determine the prefeasibility of mineral properties; it is available to the public (Smith 1992). COALVAL, another software package developed by the U.S. Bureau of Mines, can be used to evaluate coal reserves. Both programs operate as templates for well-known electronic spreadsheets for personal computers.

CHAPTER 3

Discounted Cash Flow Analysis

It can be argued that the worth of a project is the value of the project's future cash flows (or other net benefits) less the required investment (or costs). However, the time in which the investments and returns are received is also an important factor. Consideration of present and future investment and revenue streams over time forms the basis for all discounted cash flow analysis measures, such as future value, net present value, internal rate of return, and benefit:cost ratio. These methods are discussed in the following paragraphs.

CASH FLOW (CF)

Cash flow is an accounting term that represents the balance of all cash receipts minus cash operating and actual capital expenditures at the end of a year for an operating unit. It represents the amount of after-tax money that is left over or lost during a year on a cash basis. There is a cash flow for every year there is a receipt, investment, or cost for an operating unit. Cash flows may be either positive or negative; positive cash flows are reduced by debt repayment. Cash flows can be calculated either on a constant or current (inflated) dollar basis, but regardless of which basis is used, all prices, costs, and rates must be expressed in the same terms.

The forecast, or pro forma, cash flow of a project forms the basis of a number of economic evaluation methods. However, forecasting future cash flows of a project is fraught with problems. One of the major problems is forecasting product prices over the life of the project, which may be 20 years or longer. Other problems also occur in forecasting operating costs and initial capital costs, quantities produced, start-up time, and the likelihood and possible effects of such unpredictable events as floods, accidents, and political actions.

In order for cash flow analysis to be of help to host governments, all externalities should be identified, quantified to the extent possible, and either included in the cash flow or considered along with the cash flow results to reach decisions. It is the responsibility of the government, not private investors to include these externalities in the evaluation process. For example, if pollution is a concern of the host government, the social cost of pollution should be included in the project's cost and cash flow. Governments often see to it that such costs are included by mandating controls, such as specifying maximum allowable emission levels.

In addition, governments must use proper input and output prices in determining the cash flows. Market prices in developing countries can be distorted through government intervention and may not represent true prices determined by an efficient market. Under these circumstances, it may be preferable for governments to follow methodologies that are outlined by the United Nations and the World Bank and use shadow prices rather than actual prices for evaluating a project (Squire and van der Tak 1975; Ray 1984; UNIDO 1978; Dasgupta et al. 1992). Shadow prices also account for the secondary costs and benefits that may arise from a project. If distorted and unrealistic prices are used or other factors (such as exchange rates) are omitted, an improper evaluation will result and erroneous decisions are likely to be made. For further information on the evaluation of projects by governments, see the discussion of benefit:cost analysis later in this book.

The construction of a cash flow is illustrated in Table 3.1. Revenues are determined first, followed by direct operating costs, to determine operating profit. Royalties (part of direct operating costs) are usually included as charges against gross revenues to obtain net revenues. The revenues and costs account for the sales and production of all co- and by-products. Then the indirect costs, which include interest and overhead, must be determined. To determine interest, capital costs, and debt, the debt structure must first be known. Subtracting indirect operating costs from operating profit gives profit before taxes.

The next step is to calculate taxes. This is done by subtracting all allowable deductions from before-tax profit, such as depreciation and depletion, to give the tax base. Depreciation and depletion are noncash charges that are used only for the determination of taxes and have no other use in the construction of a cash flow. The significance of depreciation and depletion as noncash charges is often misunderstood because these two values are sometimes also used as proxies for actual annual capital expenditures in cost analysis.

The depreciation schedule is constructed from the actual timing, useful lives, and cost of depreciable capital items. The depreciation method depends on the tax code and goals of the governing country or state and may change over time. Almost all governments recognize at least some form of depreciation in their tax codes. Straight line depreciation, which consists of dividing the capital cost of an asset by its useful service life to determine annual allowable depreciation, is the most common depreciation method used in initial evaluations. More sophisticated depreciation methods allowed by the tax codes are used in the final evaluations. Governments allow different depreciation rates to influence investor decisions since higher rates decrease taxes and increase project profits. If a government wishes to encourage an industrial activity, it may accelerate the depreciation rate. All available depreciation options should be investigated to determine the final worth of a project.

TABLE 3.1 Sample pro forma cash flow

	Year				
	0	1	2	3	10
Sales quantity					
Product 1		100	100	100	100
Product 2 (by-product)		50	50	50	50
Net prices					
Price, product 1, $ less royalties		4.25	4.25	4.25	4.25
Price, product 2 (by-product), $ less royalties		2.00	2.00	2.00	2.00
Total net revenues, $ (= quantity × prices)		525	525	525	525
Operating costs					
Direct unit cost (product 1), $		–2.75	–2.75	–2.75	–2.75
Direct costs, $		–413	–413	–413	–413
Operating profit, $		112.5	112.5	112.5	112.5
Indirect costs					
Overhead, $		–29	–29	–29	–29
Interest, $		–5	–2	0	0
Profit before taxes, $		79	82	84	84
Depreciation, depletion					
Straight line depreciation (over 10 years), $		2	2	2	2
Percentage depletion at 10.00%, $		38	40	41	41
Cost depletion (over 10 years), $		1	1	1	1
Allowable depletion, $		38	40	41	41
Depreciation + depletion, $		–40	–42	–43	–43
Tax base, $		38	40	41	41
Taxes (46.00%), $		–18	–18	–19	–19
Capital					
Exploration, $	–5				
Development, $	–6				
Plant, property, $	–12				
Working capital, $	–149				149
Total capital, $	–172	0	0	0	0
Cash flow, $	–172	61	63	65	214
Debt/equity ratio = 0.30, $					
Debt interest (9.50%), $	–52	–52	–21	0	0
Debt repayment (max of 0.50 × cash flow), $	0	–31	–21	0	0
Cash flow from equity, $	–120	31	42	65	214

Valuation of cash flow before debt repayment (CF)	Valuation of cash flow from equity (CFE)
NPV CF for MARR of 15.00% = $186	NPV CFE for MARR of 15.00% = $195
NPV CF for MARR of 40.00% = –$13	NPV CFE for MARR of 40.00% = $6
IRR CF = 36.77%	IRR CFE = 41.64%

The depletion schedule allows a deduction from taxable income. The rationale for the deduction is that mining a mineral deposit depletes an asset that can be "renewed" only through the cost of obtaining another deposit. Depletion allowances also depend on the tax code and goals of the governing country or state and may change over time. As of 1997, the U.S. tax code allows for depletion to be calculated either on the basis of capital cost and mining rates or on the basis of a percentage of net revenue, whichever is largest, subject to certain limitations. There have been numerous proposals to eliminate the depletion allowance in the United States, which would reduce the profitability of (and thereby discourage) mining. As with depreciation, all available depletion options should be investigated to determine the final worth of a project.

Taxes can then be determined by multiplying the tax base by the tax rate. If negative taxes are due, tax statutes often allow tax credits to be carried forward to future profitable years. Parent companies can also use tax losses of their subsidiaries to offset corporate taxes in the current year.

Profit after taxes can now be determined. All that remains is to include all classes of capital investment, capital credits (undepreciated assets and working capital surplus or deficits), last-period capital expenditures (abandonment costs), and debt repayments to determine cash flow. Capital charges are subtracted during the period they are actually expended.

One category of capital often ignored is working capital, which is that portion of capital required to finance stockpiles, inventories, and the costs incurred by the lag time between expending production costs and receiving accounts receivable. The amount of working capital depends on the total operating costs and the amount of time needed between expending costs and receiving payables. The amount of time varies but can easily be 1 to 4 months of operating time. Unlike with capital spent for machinery, the investor gets working capital back when the operation closes. Working capital requirements vary with operating costs. A plant that increases production over time requires constant infusions of working capital. Yearly borrowing to meet increased working capital requirements is a characteristic of a rapidly expanding operation. Inflation will require additions to working capital even if output remains constant.

A cash flow can be determined either with or without debt. Although omitting debt, interest, and debt repayments simplifies the cash flow analysis, most large projects involve sizable quantities of debt, the effects of which must be carefully analyzed using the cash flow model.

Cash flow analysis can be constructed on either a constant or current dollar basis. Initial cash flow analysis is almost always constructed using constant dollars because including inflation greatly complicates the analysis. However, the final cash flow should be based on current dollars–after taxes and with debt accounted for–since this most closely approximates the real world.

Two final cash flows, as shown in Table 3.1 can be obtained: (1) cash flow before debt repayment and (2) cash flow from equity (after debt repayment).

While the overall cash flow of the project is of interest, in the final analysis only the equity cash flow indicates the return made on actual invested capital.

The last rows in Table 3.1 indicate net present value and internal rate of return values. These are investment merit measures based on the yearly cash flow values and the time value of money as indicated by the discount rate (MARR = minimum acceptable rate of return) and the timing of the cash flow. They are discussed in detail later in this book.

DISCOUNTED CASH FLOW

Since investors would rather receive benefits sooner than later, the value of each yearly cash flow generated over the life of a project can be adjusted for the time value of money. This is done by discounting the future value of a cash flow (FV_t) by an appropriate discount rate i and time period t to determine the present value (PV_t), or

$$PV_t = \frac{FV_t}{(1+i)^t}$$

This expression can be rewritten to show the relationship between the future yearly cash flows (CF_t) and the discounted values of the yearly cash flows (DCF_t):

$$DCF_t = \frac{CF_t}{(1+i)^t}$$

The sum of the discounted yearly cash flows gives the present value of the entire income stream at the beginning of the time period. The farther in the future a cash flow is to be received, the higher the discount factor and the lower the cash flow's present value.

DCF analysis has been a prominent technique for performing valuations and budgeting scarce capital for the past several decades. It is based on cash flow and is easily understood by engineers and accountants. However, this method of evaluation has a number of major problems (Mason and Merton 1985; Brennan 1993).

The degree to which values in the later years of a cash flow affect a DCF analysis depends on the discount rate and number of years. Figure 3.1 shows the relationships among discount rate, years, and the achieved proportion of total value for a 30-year project that consists of equal annual payments. For a discount rate of 0, achieving 90 percent of the total present value requires 27 years. For a discount rate of 25 percent, 90 percent of the total present value is achieved in the first 10 years of the project. This means that for a 25 percent discount rate, the last 20 years of the 30-year project have almost no worth from a present value perspective.

The fact that the values of cash flows in the first few years are more important on a present value basis than the later cash flows has several consequences. From a forecasting point of view, the lesser importance of future values is

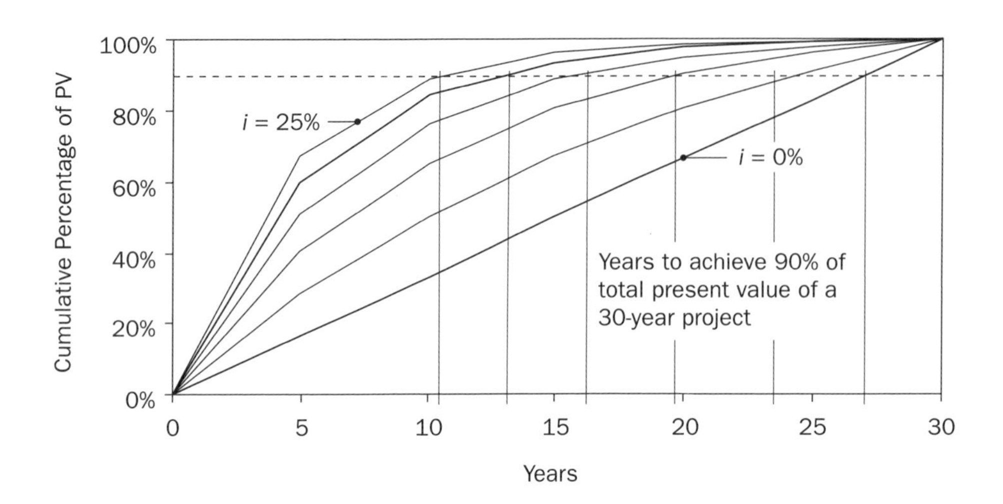

Note: Plotted lines show increasing discount rate from 0 to 25% in 5% increments

FIGURE 3.1 Influence of time and discount rate on present value

fortunate, since farther-reaching forecasts are more likely to be in error. Therefore, giving near-term forecasts higher weight may make sense. On the other hand, discounting may tend to underestimate important longer-term liabilities, such as the costs of reclamation and damages that future generations will have to bear. In addition, discounting may undervalue long-term projects because the worth of the later benefits may be understated. For example, the benefits of creating a new national park may be underestimated because future generations facing a scarcity of large preserved and accessible natural areas may place a higher value on the park than people would today.

As another example, consider the worth of a mineral property with 10 times the reserves needed to fill a 30-year contract compared to a similar reserve with only the exact amount of reserves to fill the contract. On the basis of present value of the future income streams, both properties are worth the same because the assumption is that the remaining reserves on the first property will not be mined until after the 30-year period. However, the first property would sell at a greater price than the second because the additional tonnage represents an option for future production. Option price analysis, which is discussed later in this book, may be an alternative or supplemental method for evaluating long-lived projects.

Inflation presents a particular set of problems in using DCF analysis. If inflation is assumed to be zero, all costs, prices, interest rates, and discount rates must be in constant dollars; otherwise, all values must be in current dollars. It is incorrect to mix current dollar values with constant dollar values in a single cash flow. It is also incorrect to compare the results of current dollar and constant dollar cash flows without making proper adjustments for inflation. The issue of constant versus current dollars is discussed in more detail later in this book.

DCF analysis is static in that it accounts only imperfectly for uncertainty and does not recognize the possibility of changing operations to react to changing future economic conditions. Operations can and do change according to need. For example, low prices can cause operations to shut down until prices again rise above operating costs. This limitation of DCF analysis can undervalue projects. See Lehman (1989) and the discussions of net present value and option pricing later in this book for further information concerning this weakness.

In addition to the problems in developing the yearly cash flows, DCF analysis poses problems in choosing the appropriate discount rate. While choosing the actual value of the discount rate can be complicated, the rate itself must be in either constant or current (inflated) terms consistent with the values of the cash flow. For a compilation of discount rates governments and industries use in evaluating projects, see Arizona Department of Revenue (1994).

A number of possible discount rates are commonly used in DCF analysis, although some have more credibility than others. These rates are summarized in Table 3.2. Of the discount rates identified, the opportunity cost of capital is theoretically the correct choice. Since investors have limited amounts of capital and cannot undertake all projects, they cannot go wrong by comparing the returns of a potential project with those of the next best investment alternative. However, identifying that alternative and its return is not easily accomplished.

TABLE 3.2 Summary of discount rates

Type of Rate	Description
Opportunity cost of capital	Foregone benefits that would have been received from the next best investment opportunity
Risk-free alternative	Return from a "risk-free" instrument, such as a U.S. Treasury bill. Note that even these instruments are not entirely risk-free because of risks related to the exchange rate and rate of inflation
Cost of debt	A rate based on the cost of borrowed funds
Weighted average cost of capital (WACC)	A risk-adjusted rate that weighs a firm's cost of equity and the cost of debt by the debt-to-equity ratio
Historical rate of return	A rate of return based on constant or current performance of past investments
Risk-adjusted rate of return	Any rate based on a constant or current return adjusted for project risk. The capital asset pricing method (CAPM) is often used to determine such a rate.
Hurdle rate	Any specified minimum rate of return, regardless of how derived
Social rate of return	A rate used to determine the value of social projects. This rate accounts for issues of equity and morality as well as financial aspects of an investment.
Varying discount rate over time	A rate that reflects changes in risk over time. For example, the risks can change once all capital has been recovered and a minimum rate of return achieved.
Varying discount rate by cash flow line item	A rate that reflects the difference in risk among various cash flow components. For example, working capital and inventory charges have lower risks than capital charges for new technologies.

In practice, an identified "safe" rate of return, such as interest on U.S. Treasury bills, is often used as a proxy for the opportunity cost of capital.

However, there are other views on how to appropriate the opportunity cost of capital. The capital asset pricing model (CAPM), commonly in use today in corporate finance, suggests that the weighted average cost of capital (WACC) might be the appropriate discount rate. The WACC recognizes that there is a cost of equity just as there is a cost of debt and that the debt:equity ratios of firms may vary. It also recognizes and accounts for the fact that the risk of purchasing firms' stocks will vary; thus, the WACC is a risk-adjusted discount rate. As long as the project in question has a return greater than the weighted average cost of equity and debt (i.e., greater than the cost of funds needed for the investment), it should be considered. Although the weighted average cost of capital can be calculated, its value changes over time and may depend on the types of investments made and the perceptions of lenders or stockholders. As an alternative, the cost of debt (commercial loan rate) is easily identified and is sometimes used as the discount rate, but this approach ignores the cost of equity capital, which may be different from the cost of debt. See Appendix A for details on how to calculate the WACC.

The discount rate used for any given project may be adjusted to reflect the amount of risk inherent in the project. The higher the perceived risk, the higher the risk-adjusted discount rate. While this practice is commonly employed, there is disagreement as to whether it is based on sound theoretical ground. The amount the discount rate should be adjusted for risk is often chosen in a highly subjective manner, which may lead to incorrect conclusions. Nonetheless, adjusting the discount rate for risk is the primary method used in DCF analysis to account for uncertainty.

As a project proceeds through the development stage, risks may change, which would necessitate a change in the appropriate risk-adjusted discount rate. In addition, risks may be different for various cash flows within a single project.

Since the appropriate discount rate is difficult to identify, many corporations and analysts resort to using a hurdle rate based on a number of factors. If the economic evaluation of a project fails on the basis of the hurdle rate, the project is dropped.

To avoid specifying the value of the discount rate or how it is derived, the generic term *minimum acceptable rate of return* (MARR) is sometimes used to refer to the discount rate. The discount rate and the MARR are equivalent in value and in the manner in which they are used.

Governments face the additional problem of whether to use the same discount rate as private industry or a lower, social discount rate. Since governments compete with private industry for scarce capital in the bond market, some economists argue that governments should use discount rates similar to those used by private industry. However, in that there is some risk associated with dealing with governments that private firms incur and that governments themselves do not, such as political risk, there is reason for the social discount rate to be slightly lower than the private discount rate. There are also economists who argue that the appropriate social discount rate should be very low or zero because a positive rate means that people value their own present consumption over consumption by future generations. These economists are of the opinion that such an attitude toward consumption is immoral and that any positive social discount rate is unconscionable. The question of which social discount rate to use is presently unanswered and falls in the realm of normative economics.

NET PRESENT VALUE AND INTERNAL RATE OF RETURN

Net present value and internal rate of return are the two basic measures of project feasibility for use in DCF analysis. NPV is a measure of value or of a stock of wealth, whereas IRR is a measure of the efficiency of capital use or the rate of accumulation of wealth. Both NPV and IRR are used to indicate project feasibility, but they may not result in the same ranking of a set of investment opportunities unless the IRR is correctly determined. The differences in these two evaluation criteria have caused much confusion in the evaluation profession.

Net Present Value

Net present value is the sum of the present values of all yearly cash flows less the initial investment. In short, NPV reflects the value a project appears to provide given a discount rate and a set of cash flow assumptions. Therefore, NPV is a measure of an investment's worth. The equation for NPV can be written as

$$\text{NPV} = \left[\sum_{t=1}^{n} \frac{\text{CF}_t}{(1+i)^t} \right] - I_0$$

where

CF_t = cash flow in year t

I_0 = initial investment

i = discount rate

n = total number of years for project

As an evaluation tool, NPV has many advantages. It takes into account the time value of money, and it gives a single project value for a given discount rate and set of cash flow assumptions. The larger the NPV, the richer the investors become by undertaking the project.

NPV forms the backbone of project evaluation; the NPVs of individual projects can be compared to determine comparative worth, provided each NPV is generated in a consistent manner. This means that each NPV to be compared must be determined using the same parameters, such as price assumptions, appropriately adjusted discount rates, taxation rates and consistent handling of externalities, as well as appropriate adjustments for inflation, unequal lengths of service lives and risk.

A number of important problems encountered in NPV analysis have already been discussed, including the difficulties in forecasting prices and costs and in choosing the appropriate discount rate. Small changes in cash flow assumptions, such as those for product prices, in the early years of the project can dramatically change the NPV and the investment decision. Since NPV is based on DCF analysis, it shares all the strengths and weaknesses of DCFs in general.

Internal Rate of Return

As the discount rate increases for a specific cash flow, the NPV of the cash flow necessarily decreases. IRR may be defined as that discount rate at which NPV equals zero. Alternatively, IRR may be defined as that rate that equates the initial investment with the future value of the resulting cash flows. The higher the IRR, the more profitable the project is in terms of return on invested capital. The difference between the discount rate and IRR is that the investor chooses the discount rate, whereas the characteristics of the cash flow determine the IRR. Consequently, IRR is determined internally (hence its designation as the *internal* rate of return), as compared to the discount rate for NPV, which is determined externally. The relationship between IRR and NPV can be written as

$$\mathrm{NPV} = 0 = \left[\sum_{t=1}^{n} \frac{\mathrm{CF}_t}{(1+\mathrm{IRR})^t}\right] - I_0$$

where

CF_t = cash flow in year t

I_0 = initial investment = CF_0

IRR = the discount rate that makes NPV = 0

n = total number of years for the project

While NPV and the maximization of wealth are the theoretically correct investment-ranking criteria, NPV does not indicate the return per invested dollar. IRR and other rate-of-return measurements described later, on the other hand, do give indications of the return per invested dollar. This makes IRR one of industry's most popular investment-assessing criteria. However, IRR can easily be misused.

Industry's Use of NPV and IRR

The use of NPV and IRR as merit measures by which to evaluate the feasibility of a project is well accepted by industry. A recent survey of mining companies (Bhappu and Guzman 1994) found that 55% used IRR and 40% used NPV as a merit measure. These ratios are only slightly different from two prior surveys in which 69% used IRR and 37% used NPV (Boyle and Schenck 1985) and 75% used IRR and 62% used NPV (Dougherty and Sakar 1993). IRR has clearly been the predominant merit measure used by the mining industry, although this position appears to be weakening. In addition, it appears that more companies are using both IRR and NPV as merit measures, which is appropriate since each metric presents a different perspective of value of a cash flow.

Most textbooks specify that NPV is the preferred merit measure, both for theoretical reasons related to the maximization of stockholder's wealth and for supposedly practical reasons related to problems that are thought to be associated with IRR. As mentioned in Table 3.3 and discussed in detail later in this chapter, these reasons are not persuasive. It is preferable for decision makers to have more information and additional perspective, not less. In general, decision makers will be able to make better-informed decisions if they consider both NPV and IRR.

There is considerable confusion in the literature about the definition, meaning, and use of IRR (Stermole and Stermole 1993; Au and Au 1992; and

TABLE 3.3 Summary comparison of IRR and NPV as merit measures

NPV	IRR
1. Measures the stock of wealth, which is consistent with economic theory, e.g., the maximization of utility or NPV. However, it does not tell how efficiently capital is used.	**1.** Measures the rate of wealth accumulation, or the rate of change of wealth. It indicates the efficiency of the use of capital for investments. However, it does not indicate the value of a project.
2. The size of NPV is dependent on the rate of return as well as the size of the initial investment. NPV can be made larger by increasing the size of the project.	**2.** IRR is independent of the size of the initial investment. To make IRR larger, the investment must earn a higher return.
3. Requires price and cost forecasts.	**3.** Requires price and cost forecasts.
4. Requires the choice of an external discount rate. Since choosing the proper discount rate is difficult, this requirement is often cited as a weakness of NPV that is not shared by IRR. However, this does not hold true when multiple IRRs exist.	**4.** It is commonly claimed that IRR analysis needs no discount rate except a MARR for comparison. This is true only when there are no multiple roots for IRR. When multiple IRRs exist, a specific discount rate must be chosen for comparison, just as in the case of NPV.
5. Implicitly assumes reinvestment of the yearly dividends at the MARR.	**5.** Implicitly assumes reinvestment of the yearly dividends at the IRR.
6. It is commonly claimed that NPV is a unique value that does not have the multiple-root problem that may exist with IRR. However, if multiple IRRs exist, multiple discount rates and multiple NPVs must also be examined.	**6.** Multiple IRRs may exist and, if so, may complicate the analysis. This is often incorrectly cited as a weakness of IRR that is not shared by NPV.
7. NPV analysis correctly ranks mutually exclusive projects or investments under conditions of capital scarcity.	**7.** IRR analysis correctly ranks mutually exclusive projects or investments under conditions of capital scarcity if the IRR value is correctly determined.
8. Omits the option value of production and investment flexibility.	**8.** Omits the option value of production and investment flexibility.

Lohmann 1988). The confusion centers on (1) whether IRR analysis assumes reinvestment of the yearly dividends at a rate equal to the IRR, (2) the proposed necessity of using what is called an incremental IRR to rank mutually exclusive projects, and (3) what effects of the so-called multiple-root problem may have on the usefulness of IRR as a merit measure. These problems are addressed later. For more detailed discussions of IRR, see Lohmann (1988), Beaves (1988), Hajdasinski (1984), and Appendix A.

Theoretical Conditions

DCF analysis is theoretically correct only when a number of very specific conditions are met. These theoretical conditions are seldom met in practice, which leads to numerous problems and misunderstandings. A thorough understanding of these conditions allows evaluators to correctly circumvent these problems or correctly recognize the limitations of the DCF analysis. The five necessary conditions are explained in the following five paragraphs:

1. *All input values must be known with certainty, and there must be no uncertainty or risk.* A numerical value of NPV can be correctly determined using any set of numbers, but the true value of an investment can be determined only if all input values are known with certainty. This certainty is seldom possible since future prices or costs are not exactly known. Thus, to account for risk and uncertainty, many modifications to NPV analysis are used, including sensitivity analysis, Monte Carlo simulation, and risk-adjusted discount rates. All these modifications present problems of their own. Since the presence of risk is one of the most important impediments to DCF analysis, risk is discussed later in this book.
2. *All projects to be compared must have comparable discount rates that reflect the risk-free opportunity cost of capital.* In the context of this requirement, *comparable* does not mean *identical.* Given that risks or unknowns almost always exist, risk-adjusted discount rates are used. If one project is riskier than another, it is appropriate to use different risk-adjusted discount rates when making comparisons. The opportunity cost of capital may also be different from one project to another. For example, a very large investment may restrict alternative investment opportunities more than smaller projects would. All discount rates for all projects to be compared must be in either constant or current dollars, not mixed dollars.
3. *All projects to be compared using DCF analysis must have comparable tax structures.* In other words, all projects must include similar treatments of income and other taxes. It is not correct to compare a pretax NPV with an after-tax NPV.
4. *All projects to be compared using DCF analysis must have equal equivalent economic lives if IRR is to be used as an investment criterion under conditions of capital scarcity or mutually exclusive investment.* A cursory inspection of investment opportunities may appear to suggest that, in practice, few projects have equal economic lives. However, what must be compared are the *total* cash flows that result from two or more investment opportunities over the same length of time. The returns from a 5-year investment can be correctly compared to those from a 4-year investment as long as the analysis

for the latter includes the returns from reinvestment at the opportunity cost of capital in the fifth year. Examples of the effects of unequal economic lives are given in Appendix A.

5. *All projects to be compared using DCF analysis must have identical initial investments if IRR is to be used as an investment criterion under conditions of capital scarcity or mutually exclusive investment.* Again, at first few projects may appear to have equal initial investment requirements, but what must be compared are the *total* cash flows that result from all the investment opportunities. The returns from a \$50 investment can be correctly compared to those from a \$40 investment as long as the returns from the \$10 difference, which is invested at the opportunity cost of capital, are included with the returns from the \$40 investment. In both cases, a total of \$50 is invested. Examples of the effects of unequal initial investments are given in Appendix A.

Insights Offered by a Rate-of-Return Measure

Assuming the maximization of wealth is the goal of an investor, is it always true that making decisions on the basis of NPV maximization will lead to the proper results? The simple answer is yes. However, it is quite possible for an investor to obtain more insight about an investment opportunity using IRR rather than NPV. For example, consider Figure 3.2, which shows the rate of growth of a forest and two harvesting schemes; the figure shows several cycles of growth and harvest. Trees typically experience low but increasing growth rates when they are small, increasing growth rates for a period, decreasing growth rates until they reach maturity, and negative growth rates thereafter. The problem of when to harvest the trees is the classic forester's problem described by Samuelson (1976) and many other economists since then (Lesser et al. 1997).

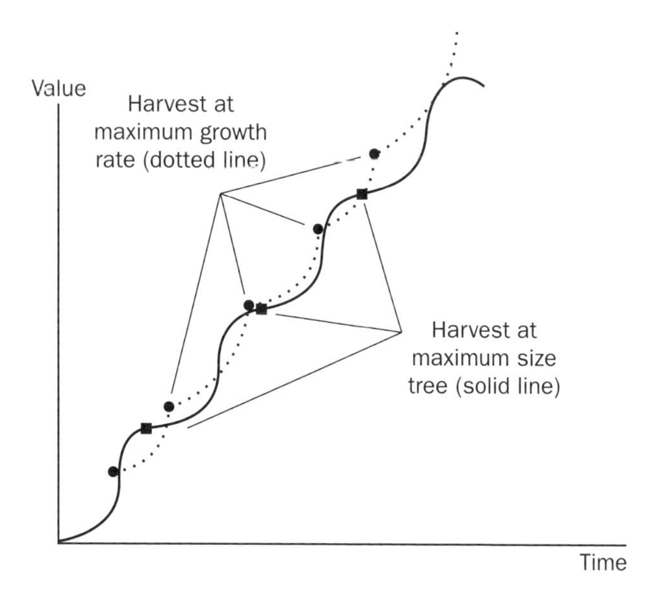

FIGURE 3.2 Forest example showing two cyclic harvesting schemes

The correct harvesting point is the time at which NPV is maximized, but the actions the forester should take based on this criterion are not obvious. Cutting the trees when they achieve maximum size is not the correct strategy. Samuelson (1976) showed that in the case of harvesting and replanting a forest—a cyclical investment—the forester can obtain greater wealth by harvesting at the time of maximum forest growth *rate*. This harvest pattern is shown by the dotted curve in Figure 3.2. In other words, the trees should be harvested when the rate of increase in wealth caused by further tree growth falls below the owner's opportunity cost of capital.

In this case, harvesting on the basis of growth rate yields the greatest wealth, the desired end product. This example shows that in some cases it may be preferable to base investment and operating decisions on the rate of growth of wealth rather than on the amount of wealth itself. This suggests that both IRR and NPV can be of importance to an investor.

IRR Reinvestment Controversy

As shown earlier, the value of IRR is determined solely on the number and amounts of the yearly cash flows, as well as the initial investment, given NPV = 0. No reinvestment of yearly dividends is explicitly factored into the IRR value itself. However, there is a controversy concerning the conditions under which the IRR can be viewed as being analogous to a compound rate of interest applied to the initial investment amount. In other words, under what conditions will the future value of the initial investment amount (if it earns compound interest at the IRR) equal the combined future value for all yearly cash flows from the project?

Mathematically, the two FV amounts will be equal only if the yearly dividends of the project's cash flows are reinvested at an interest rate equal to the IRR. Thus, in order to view the IRR as analogous to a compound interest rate on the initial investment amount, it is necessary to assume that the project's yearly dividends will be reinvested at a rate equal to the IRR. (For a numerical example, see Appendix A). If the IRR is high, it may be unreasonable to assume that the dividends can be reinvested at this rate of interest. Reinvestment opportunities at very high rates of return are rare.

Some authors have argued incorrectly that the need for this assumption makes the IRR less useful as a merit measure. However, the IRR does not need to represent a realistic reinvestment rate in order to serve as a merit measure. Whether or not the IRR method implicitly assumes a reinvestment at the IRR is largely an academic question; in either case, IRR values can be used to meaningfully compare and rank investment opportunities. The decision criterion is that an investment should be made if the IRR is greater than the MARR or, in the case of mutually independent investments or investment with capital scarcity, if the IRR is greater than the IRR of all other investment opportunities. It does not matter how large the difference between two IRR values is; it matters only that the difference is positive.

For a history of the debate, see Solomon (1956), Renshaw (1957), Dudley (1972), Grant et al. (1982), Lohmann (1988), and Beaves (1988).

Apparent Ranking Conflicts Between NPV and IRR

IRR analysis is sometimes inappropriately criticized for giving different rankings than NPV would for mutually exclusive investment opportunities. Numerous engineering textbooks state that both merit measures are useful but that a process called incremental analysis should be used to calculate an incremental, or "NPV-consistent," IRR (e.g., see Au and Au 1992; Newnan 1988; Steiner 1992; Stermole and Stermole 1993). Incremental analysis will indeed result in index values that rank projects in the same order as does NPV analysis, but it overlooks the real reason behind the difference in rankings. In addition, it negates much of the value of calculating the IRR in the first place.

The real reason there appear to be conflicts in the rankings based on NPV versus IRR is that IRR is often determined incorrectly. The following two theoretical requirements for DCF analysis (as discussed earlier) are often violated in IRR calculations: (1) that all investments to be compared must have equal economic lives, and (2) that all projects to be compared must have equal initial investments. In practice, violation of these two conditions is the rule rather than the exception. Making the adjustments discussed earlier to account for these requirements will allow IRR values to be calculated correctly.

The idea of determining NPV-consistent IRR values is flawed in that it essentially negates IRR's usefulness as a separately determined merit measure. It yields values that simply mirror the NPV results rather than providing additional information. Investors frequently have multiple decision criteria (e.g., see Baumol 1965; Evans 1984; Walls 1995; Walls and Eggert 1996). Thus, evaluators benefit from having both total-value *and* return-on-investment merit measures. There is no all-encompassing single merit measure that can relieve investors of having to make decisions. In the case of mutually exclusive projects, investors should consider both NPV and IRR in light of their own personal viewpoints and goals.

For examples concerning ranking conflicts between NPV and IRR and the determination of incremental IRR, see Appendix A.

The Multiple-Root Problem

In cases when a project has large capital expenditures both at start-up and in later years, it is possible for more than one IRR to exist for a given cash flow. Consider Figure 3.3, in which the cash flows start out negative as a result of the initial investment but then turn positive; a second investment then causes the cash flows to turn negative again before recovering and becoming positive. As a consequence, there are three specific discount rates at which NPV = 0. That is, there are three roots to the equation that defines IRR and, correspondingly, three values of IRR. Figure 3.3 also shows that in cases where multiple roots exist, higher discount rates will not necessarily yield lower NPVs.

The use of IRR as a merit measure has sometimes been criticized because the existence of multiple roots does not allow the correct IRR to be identified

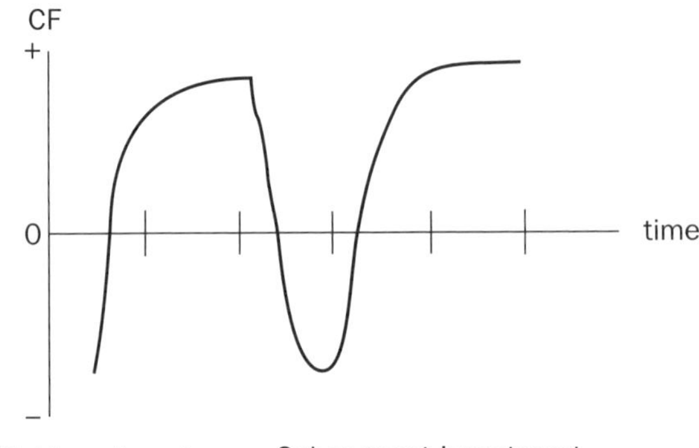

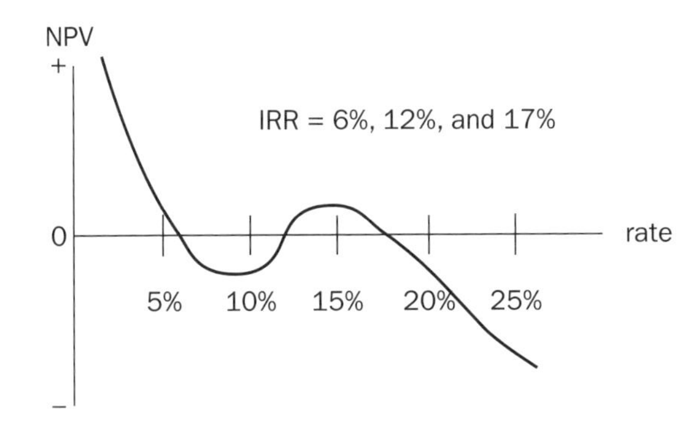

FIGURE 3.3 Multiple investments and IRR values

immediately for ranking the project in question. Some engineering economy textbooks incorrectly suggest that NPV—because it does not lead to multiple values—is a superior measure.

In practice, however, a value for MARR is chosen before NPV can be determined. NPV will be either positive or negative depending on the choice of MARR. (Recall from earlier that the MARR and discount rate are equivalent.) For instance, as Figure 3.3 shows, a discount rate of 10% yields a negative NPV, whereas a discount rate of 15% yields a positive NPV. Thus, NPV depends on the discount rate (or MARR) just as IRR does. In other words, the multiple-root problem does not represent a disadvantage of IRR relative to NPV. All merit measures for DCF analysis are affected similarly when multiple roots exist.

Multiple IRRs are commonly thought to exist only for unconventional cash flows. A single IRR does, of course, simplify the analysis. However, it is not unusual for a project to experience negative cash flows as a result of the staged and anticipated need for additional capital after start-up. For the purposes of judging IRR as a merit measure, the existence of multiple IRR values should be considered the general case rather than the exception. Whatever applies to the more general case of multiple IRR values also applies to the simpler case of a single IRR. Fortunately, investments involving multiple infusions of capital usually occur as a series of independent projects and decisions, so cases of multiple IRR values are less common than would otherwise be expected.

For numerical examples demonstrating the multiple-root problem, see Appendix A.

GROWTH RATE OF RETURN, PRESENT VALUE RATIO, AND OVERALL RATE OF RETURN

Three other commonly used merit measures are growth rate of return (GRR), present value ratio (PVR), and overall rate of return (ORR). These merit measures are closely related to NPV and IRR and share the same strengths and weaknesses. For example, since NPV and IRR have problems associated with multiple roots (as described earlier), all merit measures derived from NPV or IRR have the same problems. The characteristics and uses of all of these measures are often confused.

GRR is a rate of return that includes a return on dividends from the initial investment discounted at i = MARR plus a return from the reinvested dividends at an external rate i'. Thus, GRR is identical to IRR except that the reinvestment rate may not necessarily equal IRR. If the reinvestment rate i' does equal IRR, then GRR and IRR will be the same. GRR analysis will lead to the same ranking as does IRR as long as a common i = MARR and external i' are used for all projects. A numerical example involving GRR is given in Appendix A.

PVR is a measure related to NPV; it is represented by

$$\text{PVR} = \frac{\sum_{t=1}^{n} \text{CF}_t(1+i)^{-t}}{I_0}$$

where

t = the year

n = total number of years

CF_t = cash flow for year t

I_0 = initial cash flow

i = the discount rate

The decision criteria is to accept the opportunity if PVR > 1. This ratio has the same multiple-root characteristics as any other rate-of-return measure, such as IRR, although this is not obvious. PVR gives the same ranking as does ORR, but it will not necessarily give the same ranking as does IRR unless IRR is calculated correctly to account for the total amount and time of investment, as discussed earlier.

ORR is a merit measure that is related to the PVR where ORR = 1 – PVR. ORR can also be calculated as

$$\text{ORR} = \frac{\left[\sum_{t=1}^{n} \text{CF}_t(1+i)^{-t}\right] - I_0}{I_0}$$

where

t = the year

n = total number of years

CF_t = cash flow for year t

I_0 = initial cash flow

i = the discount rate

ORR is an unambiguous measure of return on invested capital. The decision criterion is to accept the opportunity if ORR > 0. ORR ranks projects in the same order as PVR and has the same difficulties with multiple rates of return as does IRR, but it may not rank projects in the same order as does IRR unless IRR is calculated correctly to account for the total amount and time of investment, as discussed earlier.

BENEFIT:COST ANALYSIS

Benefit:cost (BC) analysis is another method used, usually by governments, to evaluate projects. It is often synonymous with the analysis of social benefits and costs of a project, although it can be used to evaluate projects from other than a social perspective.

BC analysis has an extensive literature that started in the early 1970s because of massive investments in the public sectors of developing countries, as well as interest on the part of the World Bank and the United National Industrial Development Organization (UNIDO) (UNIDO 1972; Squire and van der Tak 1975; UNIDO 1978; Ray 1984; Gittinger 1982).

Little and Mirrlees (1990) and Dasgupta et al. (1992) discuss the usefulness of BC analysis and point out the need to use shadow prices to determine the social value of projects when actual prices are distorted by government policies. If distorted prices were used in a BC analysis, a private investment that yields low or negative economic returns for the country as a whole might be identified as justifiable. Another way to approach the inclusion of secondary benefits and costs is through the use of input/output analysis.

If the value of the benefits of a project exceeds the costs, the BC ratio will be greater than 1, which indicates the project is worth undertaking. In the case of mutually exclusive investments, the decision criterion is to choose the project with the highest BC ratio greater than 1. Since a BC ratio is a measure of the rate of growth rather than value, it has many of the same characteristics as IRR. If both benefits and costs are discounted at the same discount rate, BC ratios will give the same ranking of independent and mutually exclusive projects as do IRR values. As with IRR, BC ratios will give the same ranking results as NPVs as long as the ratios are calculated correctly to account for unequal project lives or unequal initial investments among projects.

Although NPV analysis gives the same results as correctly and consistently calculated BC ratios with less trouble, many governments often specify the incremental BC ratio as the evaluation measure that must be used. There are numerous ways to define and calculate the benefits and costs, and individual agencies may have different accepted practices. One problem with BC ratios is that the value of a ratio will be different depending on whether a cost is entered as a cost in the numerator or as a negative benefit in the denominator. Two analysts using the same data but applying it differently could arrive at conflicting conclusions. Therefore, BC ratios of several projects cannot be reliably compared unless all ratios are determined in the same manner. For additional discussion on the practical problems and history of BC analysis, see Steiner (1992) and Little and Mirrlees (1990).

HOSKOLD FORMULA

One of the first valuation methods designed to determine the value of mineral projects utilized the Hoskold formula. This method was used prior to the advent of conventional DCF analysis and corporate income taxes. Although corporations today do not generally use the Hoskold formula to determine mineral property values, governments sometimes use it to determine value for the purposes of taxation. The concepts of the formula are of current interest, and modifications of the original Hoskold formula may have useful applications (de la Cruz 1980). The Hoskold formula is expressed as follows:

$$P = A\left[\frac{i}{(1+i)^n - 1} + i'\right]^{1}$$

where

P = sinking fund present value

A = uniform annual net profit

i = "safe" rate of return

i' = speculative (risk) rate of return

n = number of years or mine life

As the equation shows, the Hoskold method assumes there are two applications of cash flow derived from a mining project. The first is to repay invested capital by investing a portion of the cash flow into a sinking fund at a safe rate of interest. The purpose of the sinking fund is to replace the property once depletion has occurred. Creation of the sinking fund assumes that a mining operation would wish to stay in the mining business over the long run and that another property could be obtained at the same cost as the original. The second application is to pay the investor a higher rate of return for undertaking a risky mining project.

Although the concept of recognizing that a cash flow may have different risk components is valid, the application of funds by mining companies as assumed by this equation is not. A number of inaccuracies in the Hoskold formula are illustrated by de la Cruz (1980). Mining companies invest in projects that offer the highest return; they do not invest portions of cash flow at a riskless return to provide funds to replace depleted deposits. Therefore, the Hoskold formula uses a lower discount rate than a corporation would actually use, which results in an overvaluing of the project.

Although investors do not generally use the Hoskold formula to determine project worth, institutions and governments sometimes use it for specific applications, such as the assessment of a mining property in which periodic contributions must be made to ensure funds are available for closure. In this case, a company may be required to invest in a low-risk sinking fund to provide sufficient funds to correct environmental problems upon closure of the mine. Alternatively, the government may require the company to post a bond to ensure that reclamation will be properly completed.

For additional information on the use and modification of the Hoskold formula, see Dran and McCarl (1974), Parks (1957), and de la Cruz (1980).

CHAPTER 4

Accounting for Inflation and Varying Demand

In any project evaluation, it is necessary—but difficult—to be consistent in the use of constant or current dollars for making forecasts. In addition, evaluators face a great challenge in forecasting prices given the unknowns in the supply of and demand for minerals. These issues together account for a large proportion of the errors typically made in an evaluation.

CONSTANT OR CURRENT DOLLARS?

Inflation causes money that will be spent or received in the future to be worth less than money in the present. Since inflation does exist and varies greatly depending on the economic cycle, a question arises as to whether evaluations should be based on constant or current dollars. The answer depends on the type and quality of information available and desired. Both types of analysis will give similar, but not equal, answers.

Constant dollar analysis is more frequently conducted because evaluators usually do not know enough about inflation to be able to forecast it on a longer-term basis. However, if inflation can be forecast, current dollar analysis gives more reliable results; it allows evaluators to account for the effects of tax deductions and taxes and, perhaps, inflation rates that vary with time. However, given the difficulty in forecasting any inflation rate, the practical worth of using multiple or changing inflation rates to determine NPV is uncertain.

Care must be taken to express all variables in an analysis consistently in either constant or current dollars. If inflation is assumed to be zero, all costs, prices, interest rates, and discount rates must be in constant dollars. If inflation is assumed to be present, all values must be in current dollars. It is incorrect to mix current dollar values with constant dollar values in a single cash flow. It is also incorrect to compare the results of a current dollar cash flow to a constant dollar cash flow without making proper adjustments for the effect of inflation on the results.

Analysts must be sure not to confuse changes in costs or prices due to inflation with real changes in relative prices or costs. Inflation is caused by a country's macroeconomic policy and is beyond the control of an individual investor. On the other hand, real price changes, such as those caused by meeting increased emission standards or those due to imbalances in supply and demand, are not caused by inflation. Real changes in relative values over time should be treated on a constant dollar basis.

It is beneficial to conduct all forecasts and initial analyses in constant dollars to maintain a sense of the relative sizes of the numbers. Table 4.1 gives examples of the difference between cost and price changes due to real changes versus a combination of real changes plus inflation. For example, it might be expected that the real price of labor will increase 1.0% per year, which means the wage rate will increase from $35,000 to $38,662 in 10 years, as shown in columns 2 and 4 of Table 4.1. The real cost to the investor thus increases 1.0% per year. If inflation is at 3.0% per year, the overall wage rate will increase from $35,000 to $51,958 in 10 years, as shown in column 6. However, since all other costs and prices increase at the general rate of inflation, the real cost to the investor still increases only 1.0% per year.

FORECASTING SUPPLY, DEMAND, AND PRICES

The first items to be entered into a pro forma cash flow are prices and quantities of the products so that revenues can be determined. This step represents the first of several problems involved in conducting discounted cash flow analysis. If a 10- or 20-year cash flow is to be constructed, prices of all products must be determined for the full period. This is a formidable task since it is difficult to forecast prices of most commodities even a year in advance.

TABLE 4.1 Example of constant and current dollar forecasts

Item	Today's actual values ($)	Real percent change per year, i	Constant dollar forecast 10 years from today, no inflation ($)	Real percent change plus 3.0% annual inflation, i' *	Forecast 10 years from today, 3% annual inflation, i'' ($)
Labor, $/year	35,000	1.0	38,662	4.03	51,958
Oil, $/bbl	18.00	5.0	29.32	8.15	39.40
Gold, $/oz	350	3.0	470	6.09	632
Copper, $/lb	2.00	−.5	1.90	2.15	2.47

*The discount rates can be determined by

$$i = \frac{(1 + i')}{(1 + i'')} - 1$$

where

i = real rate, or rate without inflation

i' = nominal or inflated rate

i'' = the rate of inflation = 3.0%

To illustrate the difficulty in forecasting prices, consider Figure 4.1 which shows price behavior for a commodity over a period of time. Since a clear trend exists, it is tempting to simply extend the path along the same trend. However, basing a multimillion dollar capital expenditure on such forecast results would be foolish. More information is required than the simply historical price trend.

Prices are determined when buyers and sellers agree on a transaction. The schedule by which buyers and sellers are willing to buy and sell can be represented by demand and supply functions or curves. These functions are often depicted as downward-sloping demand curves and upward-sloping supply curves, as shown in Figure 4.2. This type of figure is familiar to anyone with a very basic knowledge of economics.

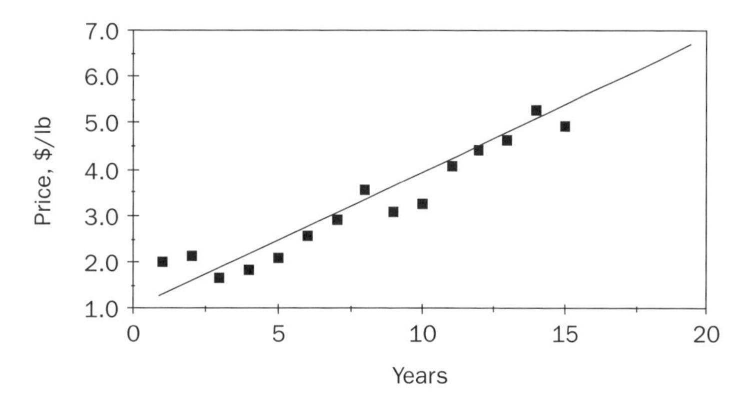

Note: Points represent actual values; solid line represents overall trend, with simple extrapolation into the future.

FIGURE 4.1 Sample historical price trend

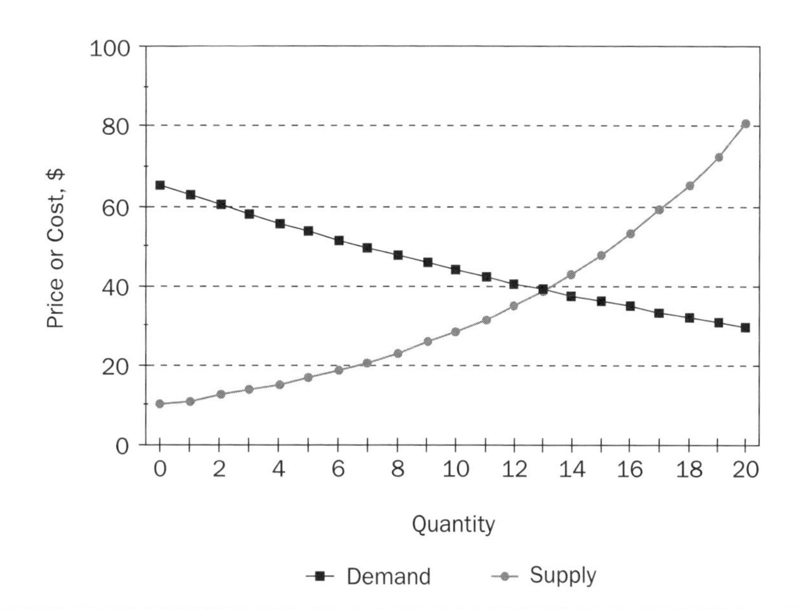

FIGURE 4.2 Supply and demand functions and price

The intersection of the two curves represents that price at which buyers are willing to buy and sellers are willing to sell. Therefore, Figure 4.1 actually represents a series of intersections of supply and demand curves. In practice, it may not be possible to tell anything more about the shape of the supply or demand functions than identifying where they intersect. However, economists have identified a number of factors that determine the supply and demand of a good. That is, there are identified factors that determine the shape and position of the supply curve relative to the shape and position of the demand curve:

Factors affecting supply	Factors affecting demand
▪ price of labor	▪ price of the good
▪ price of energy	▪ prices of all substitute goods
▪ price of materials	▪ technology
▪ quantity of labor needed	▪ consumer tastes
▪ quantity of energy needed	▪ consumer disposable income
▪ quantity of materials needed	▪ other factors
▪ price of capital	
▪ quantity of capital needed	
▪ supply or production technology	
▪ other factors	

Forecasting prices over a number of years means forecasting the many individual components of the supply and demand functions. This complexity makes it easy to understand why prices are so difficult to forecast. Economists have attempted to correlate supply and demand factors with prices in an effort to formulate long-term price forecasts, but even the most sophisticated efforts have enjoyed little success.

Other factors that affect the value of a project are capital and operating costs. Analysts spend a great deal of time determining what these costs will be for new projects. The engineering profession has become very good at determining what future costs will be from operations not yet constructed. However, as discussed in a previous chapter, net present values of anticipated cash flows are almost always more sensitive to changes in prices than to changes in operating or capital costs.

CONCLUSION

At the very beginning of a discounted cash flow analysis, analysts encounter serious difficulties for which there are few answers and no good ones. The potential errors caused by the inability to forecast prices overshadow many of the other items in a cash flow analysis. Analysts and decision makers should never lose sight of this tenuous beginning, say, when the computer determines a rate of return to several decimal places or when the returns of two mutually exclusive projects differ by only small amounts.

CHAPTER 5

Incorporation of Risk in Project Analysis

Many factors, prices, costs, schedules, and quantities may not be known with much certainty for any given project. Therefore, decision makers must try to understand the consequences of not knowing these values and plan operations accordingly. Methods used to analyze cash flows to determine the effects of these uncertainties include sensitivity, scenario, Monte Carlo simulation, certainty equivalence, Bayesian, and political risk analyses.

SENSITIVITY ANALYSIS

Sensitivity analysis is simply the process of varying one or more factors to see what the variance does to the value of the project. While sensitivity analysis contributes to understanding the effects of uncertainty, it does not give a project value adjusted for the perceived uncertainty.

One of the great values of sensitivity analysis is that it identifies those factors that most greatly affect a project's economics. This allows evaluators to gather additional data in a more efficient manner. For example, there is no point in improving beneficiation cost data if the variations in such data are overshadowed by great expected price variations for a particular project.

A simple tool that can be used to help in sensitivity analysis is known as a spider diagram. An example of such a diagram for a hypothetical investment opportunity is shown in Figure 5.1. A spider diagram can be used to determine which factors contribute most to variances in NPV or IRR. The example shown in Figure 5.1 suggests that changes in commodity price (i.e., revenue) affect NPV more than changes in either capital or operating costs. This is the usual case because revenue includes all sources of income, whereas operating costs and capital represent only portions of total costs. Therefore, a percentage change in the smaller operating or capital costs does not have as much effect on NPV as a percentage change in the larger total revenue figure.

Another important factor affecting the valuation results is the timing of the investment relative to price cycles. Sensitivity analysis can also be performed

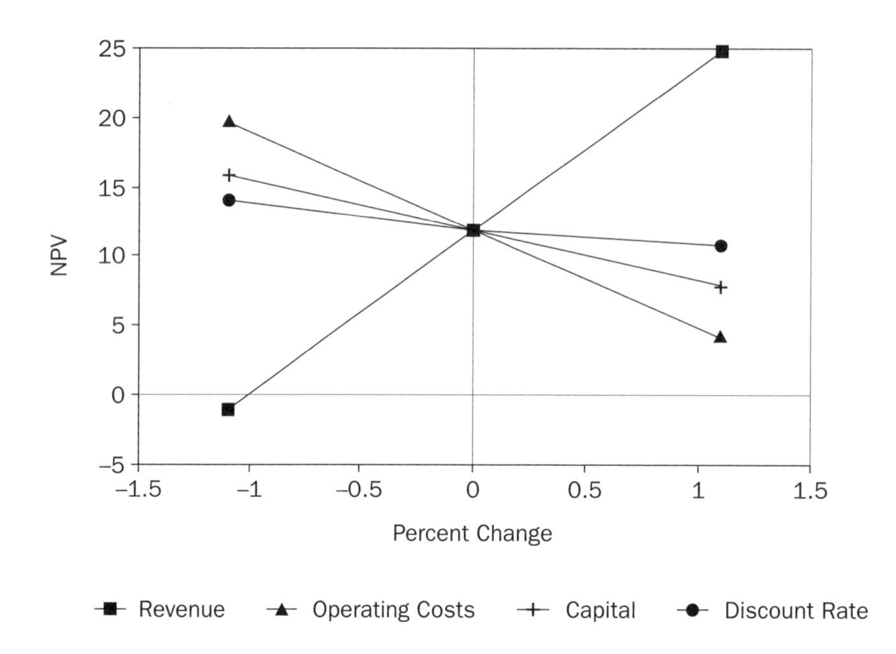

FIGURE 5.1 Example spider diagram showing relative importance of cash flow components

to provide an understanding of the importance of price cycles on project value. A more detailed discussion of this important aspect is presented in Chapter 6.

SCENARIO ANALYSIS

Decision makers understand the uncertainty created by multiple combinations of factor values; as a result, they sometimes investigate the results of scenarios in which combinations of variables are changed. This type of approach is known as scenario analysis. Consider first a "base" case as shown in the simplified pro forma cash flow for a potential investment in Table 5.1. The cash flow consists of a single investment and 10 years of equal revenues and costs on a constant dollar basis. If the minimum acceptable rate of return of the potential investor on a risk-adjusted basis is 10%, the NPV of the project is $5,692. For the sake of simplicity, assume that the risk-free discount rate is 6.00%, which implies that the risk component of the discount rate is about 4%.

In this case, a risk-adjusted discount rate is used to deflate the values of future incomes. While there are numerous theoretical problems in using such a rate to account for risk, the immediate problem is how to choose a risk factor that accurately represents the inherent risk actually present. A subjective assessment of the risk can be made and a subsequent risk factor chosen, but this procedure does not build confidence in the outcome of the investment decision because the risk factor may not be correct. If the true risk is much higher, the appropriate higher risk-adjusted discount rate might result in a negative NPV for the project, which would cause the project to be rejected. Although it is not desirable to pass up a good project, it is even more important not to lose large amounts of money. However, rejection may not be fully justified.

The problem the decision maker faces is caused by insufficient information to make an informed decision. The single pro forma cash flow and the NPV figure do not give any indication of the variances that may be expected in the factor input values or what effect these variances may have on project worth. One way to identify and quantify these unknowns is to construct scenarios involving the expected ranges of input variables. An example of the construction and results of three scenarios–showing the optimistic, base case, and pessimistic outcomes–is shown in Table 5.2.

The base case is constructed from the "best" estimates of the project parameters, and the resulting NPV is often–though incorrectly–thought of as the "expected value" of the project. (In a statistical sense, this base case is probably not the expected case. The expected case can be determined only through probabilistic analysis, as the distribution of the input factors has not been considered in the determination of the base case.) The pessimistic case shows the results of what happens when everything goes poorly, and the optimistic case shows what happens when everything goes well. As Table 5.2 shows, variables not known with certainty include price, quantity sold, wage rate, energy price, materials cost, number of workers, amount of energy required, amount of materials required, and the amount of capital needed. With a MARR of 10%, the NPVs of the three scenarios for this example are $3,265,000, $5,692, and -$993,487, respectively. The question now facing the decision maker is whether the project should be undertaken. While it is comforting that the "expected" NPV is positive and the optimistic NPV shows the possibility of much greater than minimum profits, the possibility of losing nearly $1 million on a $450,000 investment is disconcerting, to say the least.

TABLE 5.1 Hypothetical base case pro forma cash flow

	Year			
	0	1	2	10
Product price, $		2.50	2.50	2.50
Quantity		60,000	60,000	60,000
Total revenues, $		150,000	150,000	150,000
Wage rate, $		15,000	15,000	15,000
Energy price, $		2.50	2.50	2.50
Materials cost, $		3.00	3.00	3.00
Number of workers		5	5	5
Units of energy		500	500	500
Units of materials		8,000	8,000	8,000
Total operating cost, $		100,250	100,250	100,250
Capital cost, $	300,000			
Cash flow, $	(300,000)	49,750	49,750	49,750

NPV for MARR of 10.00% = $5,692

IRR = 10.44%

An investor using scenario analysis in this manner may well make an incorrect decision. The analysis in Table 5.2 gives the dangerous and erroneous illusion that the median or expected value of the project is $5,692, that the most the project will make is $3,256,000, and that the most the project will lose is $993,487. All three of these conclusions are misleading.

First, it is unlikely that all costs and quantities for each of the scenarios are associated with similar probabilities of occurrence. In the optimistic case, if the subjective probability that the energy price will be no greater than $2.00 is not at least similar to the subjective probability that the capital cost will be no greater than $250,000, then the entire scenario analysis is fatally flawed and useless.

Even if all values for each of the scenarios are associated with similar probabilities, the results of the analysis are still misleading. Probability theory shows that the joint probability of all prices being high and all costs being low at the same time is extremely low. More likely, some prices will be higher than expected, some lower, and some about as expected. It makes no sense to base business decisions on the occurrence of events that are highly unlikely to happen.

Furthermore, even the base case "expected" value may be highly misleading. As will be shown in the next section on probabilistic analysis, the true expected value must be determined in a manner different from simply constructing a cash flow from values thought "likely" to occur.

TABLE 5.2 Different scenarios using base case as starting point

	Yearly Cash Flows (Life = 10 years)		
	Optimistic	Base	Pessimistic
Product price, $	8.00	2.50	2.00
Quantity	80,000	60,000	40,000
Total revenues, $	640,000	150,000	80,000
Wage rate, $	12,000	15,000	16,000
Energy price, $	2.00	2.50	3.50
Materials cost, $	2.75	3.00	4.00
Number of workers	4	5	8
Units of energy	350	500	700
Units of materials	7,000	8,000	9,500
Total operating cost, $	67,950	100,250	168,450
Initial capital cost, $	250,000	300,000	450,000
Cash flow, $	572,050	49,750	(88,450)
NPV for MARR = 10.00%, $	3,265,000	5,692	(993,487)
Payback, years	0.5	6.0	Never

Scenario analysis is a widely used but potentially very dangerous tool for the decision maker. The NPVs for each of the scenarios are mathematically consistent, which gives a sense of respectability to the process, but the conclusions a decision maker might reach from these results may well be erroneous. Scenario analysis can be useful but should be used only with extreme caution.

PROBABILISTIC ANALYSIS AND MONTE CARLO SIMULATION

Probabilistic analysis can be thought of as the ultimate form of scenario analysis in that all possible cases are considered simultaneously. The input for the analysis consists of a distribution of values for each variable in a cash flow analysis. In other words, for each variable used in a cash flow, a range of values and their probabilities of occurrence are used as inputs instead of a single value as in scenario analysis. Since inputs are probabilistic, most of the risk inherent in the project is reflected in the range of input variables. Therefore, the discount rate used in probabilistic evaluation methods must reflect the risk captured in the cash flow itself. The risk component in a risk-adjusted discount rate decreases in proportion to the amount of risk expressed in the probabilistic range of input values. If all risks are totally expressed in the probabilistic determination of the range in values of the inputs, a riskless discount rate must be used. This is in contrast to a risk-adjusted discount rate that may be used for nonprobabilistic evaluation methods, such as scenario analysis. A computer program using a method called Monte Carlo simulation can then be used to generate hundreds of variations of cash flows (i.e., hundreds of scenarios) and NPVs for an individual project. The statistical distribution of the NPVs is then analyzed to determine the worth of the investment opportunity. The mean of the NPVs obtained by Monte Carlo simulation represents the statistically defined expected value of the project.

The Procedure

The Monte Carlo simulation procedure starts with the construction of a base cash flow—for example, as shown in Table 5.1—and the identification of the ranges and distributions of the important variables. For the sake of discussion here, the values shown in Table 5.2 can be taken to define the ranges of inputs for this example problem. Probability distribution functions must next be determined for each range of values. Although any suitable type of distribution can be used in the probabilistic simulation process, consider simple triangular distributions for each of the variables shown in Table 5.2. In this manner, the lowest and highest values are considered to have low probabilities of occurring (say, less than 5%) and the base values the highest probabilities of occurring.

The use of probabilistic inputs for the variables accounts for the risk involved with the uncertain values of these variables. The discount rate used in Monte Carlo simulation should thus reflect this internalization of risk. Consequently, the appropriate discount rate should include a risk factor less than the 4% assumed in the scenario analysis example. However, for instructional purposes, a discount rate of 10% will be used in this Monte Carlo simulation to illustrate the difference in NPV due solely to the probabilistic characteristics of the simulation. For a discussion of the proper use of risk-free discount rates in Monte Carlo simulation, see Davis (1995a, 1995b) and Pindred (1995).

The next step is to estimate possible correlations among the input variables and to make the appropriate entries in the probabilistic program. For the example given in Tables 5.1 and 5.2, assume that all variables are independent and no correlation exists. The program then proceeds to choose a random value from the distribution of each of the input variables and to determine an NPV from the chosen values. This procedure is reiterated a sufficient number of times so that a statistically large number of different combinations of values are selected and a statistically large number of NPVs are generated. The larger the number of variables to be sampled, the larger the number of iterations must be. For this example, 300 iterations were run, resulting in the determination of 300 NPVs.

An example of the triangular distribution of the product price is shown in Figure 5.2. As the figure shows, the triangular distribution is not symmetrical but rather skewed toward the right. The expected range varies from $2.00 to nearly $8.00, with the modal value, or the one that occurs with the greatest frequency, of about $2.50. However, the calculated mean, or expected value, is $4.17, which indicates that the price is more likely to be above the modal value of $2.50 than below it.

The relationship between price and probability is more clearly shown on a cumulative distribution plot, as shown in Figure 5.3. Because of the skewed price distribution, there is a probability of greater than 80% that the actual price will be greater than $2.50. This observation is entirely missed in the scenario analysis of Table 5.2. The figure also shows there is a 50% chance that the price will be either above or below $3.90.

Similar plots can be made for the distributions of the other input variables, as shown in Figures 5.4 and 5.5. Figure 5.4 shows the distribution for quantity produced and sold. This triangular distribution is symmetrical and has the same property as a normal distribution in that the value that occurs the most is also the mean value. Figure 5.5 illustrates the skewed triangular distribution for capital cost. A distribution for estimated capital expenditures is typically skewed since there is almost always a greater probability of the costs being higher than being lower.

The distribution of the NPVs (based on 300 iterations) is shown in Figure 5.6. It appears to be bimodal (i.e., to have two distinct peaks), but this characteristic is actually related to the number of iterations and number of observations included for each data bar rather than to a systematic relationship among the input variables.

The plot of the cumulative distribution of NPVs, as shown in Figure 5.7, contains much valuable information for making investment decisions. This plot clearly shows there is only a small probability (less than 5%) that the value of the project will be less than $200,000 or more than $800,000, even though the NPV distribution is skewed toward the right at higher values. It also shows that the mean value will be about $530,000. The mean value of expected

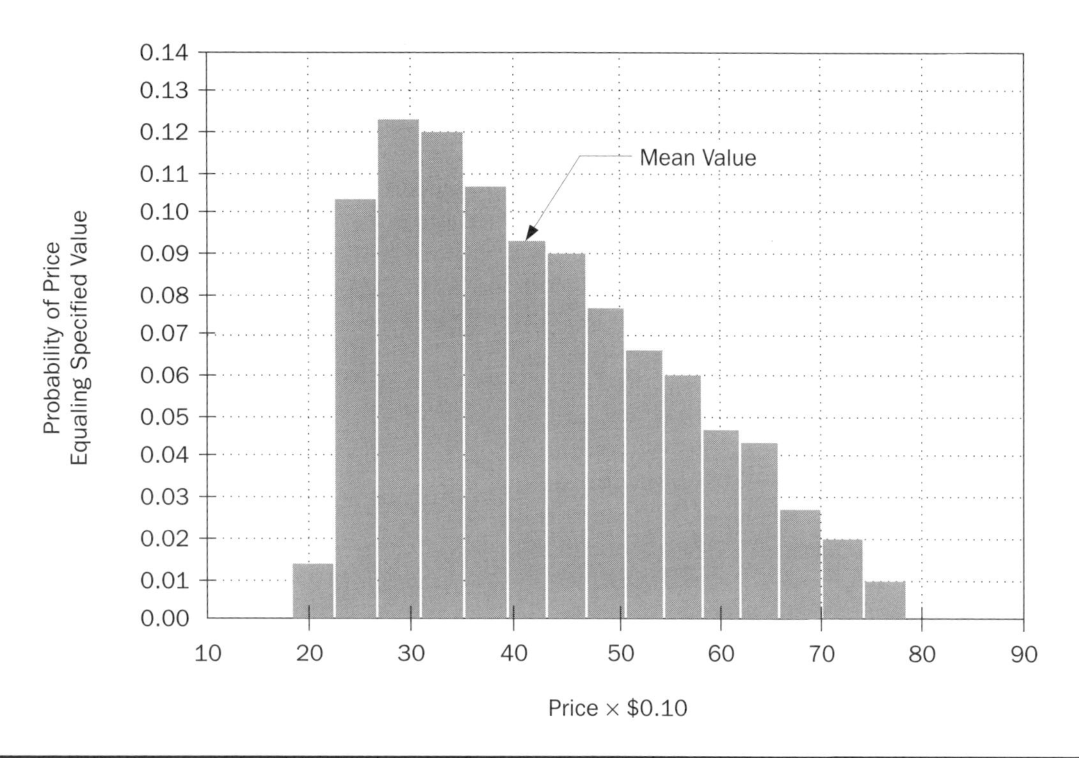

FIGURE 5.2 Distribution for price

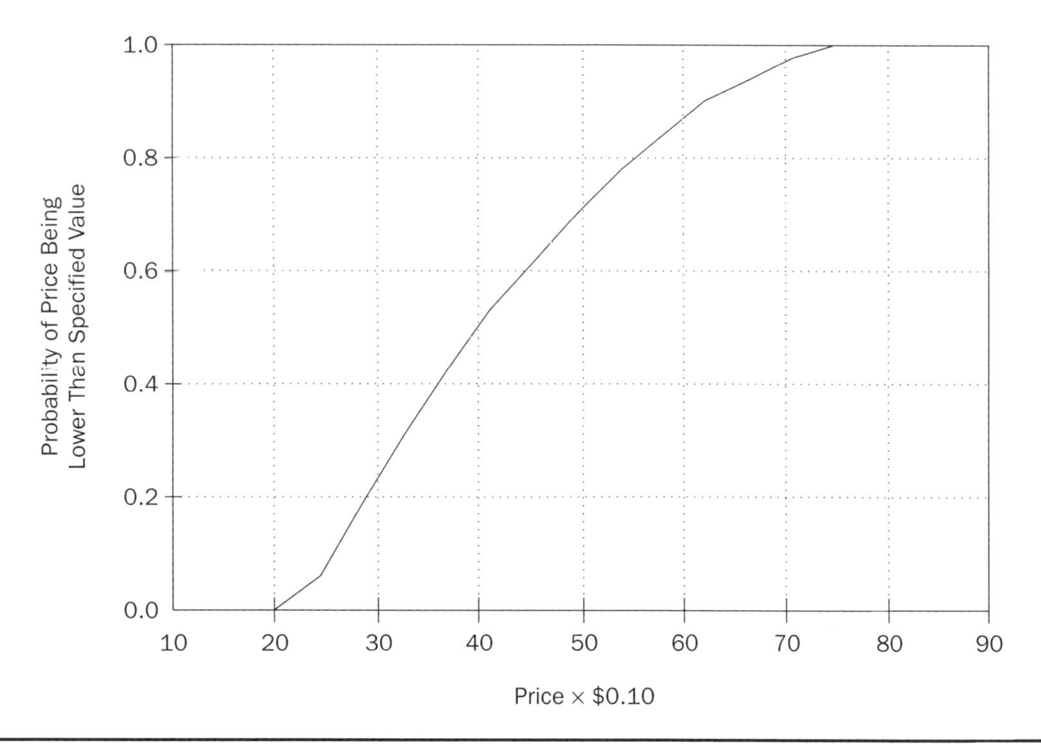

FIGURE 5.3 Cumulative distribution for price

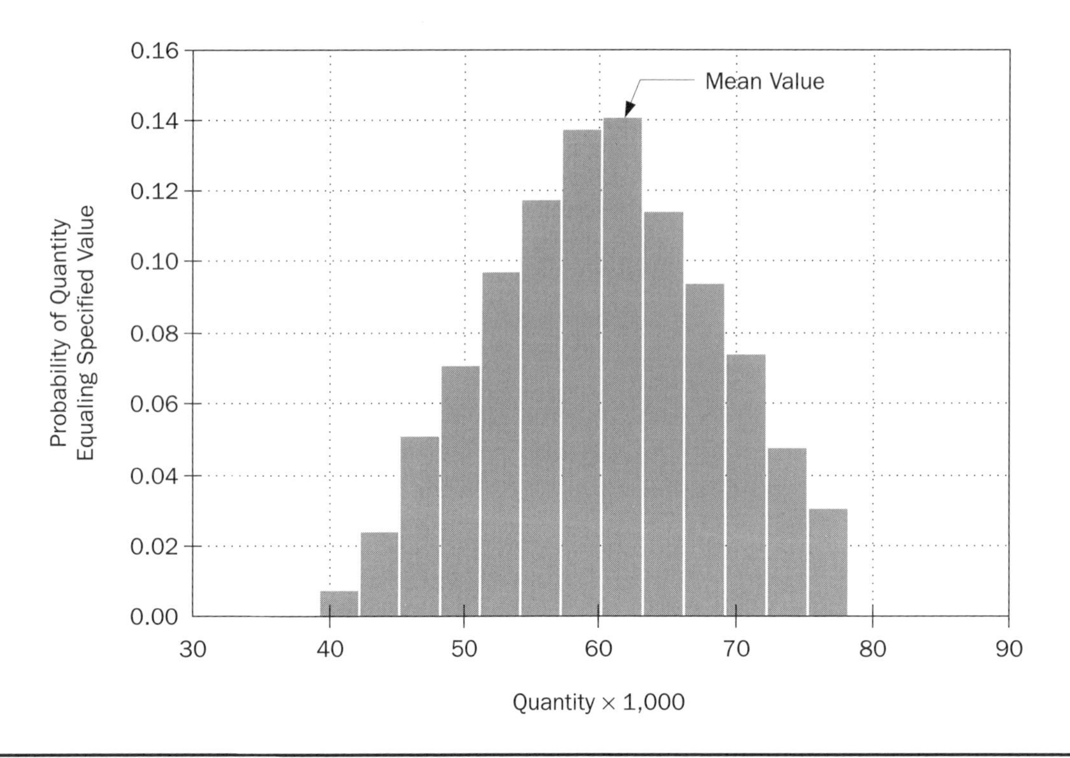

FIGURE 5.4 Distribution for quantity

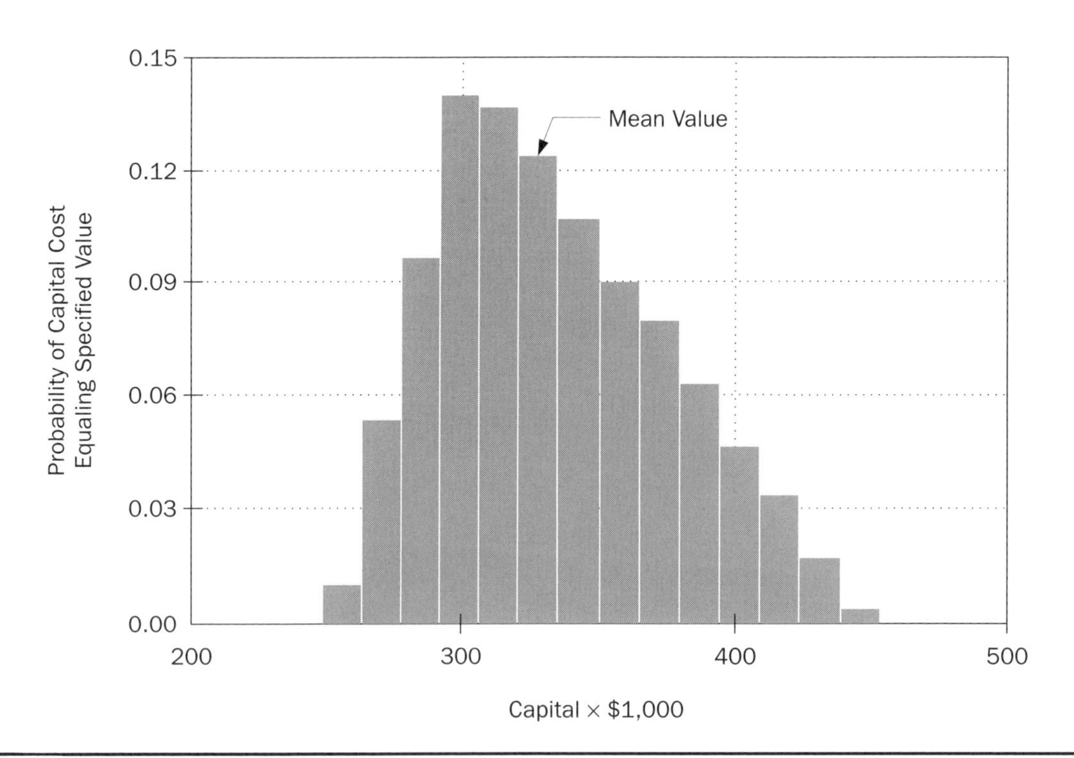

FIGURE 5.5 Distribution for capital cost

outcomes represents the expected value of the investment. A display of the mean values of the inputs and the resulting mean NPV is shown in Table 5.3.

Values shown in Table 5.3 and Figure 5.7 contrast sharply with those obtained from the cash flow analysis shown in Table 5.1 and the scenario analysis shown in Table 5.2. The "expected" NPV of $5,692 shown in Tables 5.1 and 5.2 sharply contrasts with the calculated expected value of $532,180 obtained through probabilistic simulation. This difference is caused by recognizing that there is a distribution for the input values. The values in Table 5.1 do not take into account the probabilistic distributions of the input values. While a price of $2.50 may be the single value that occurs the most frequently, as shown in Figure 5.2, the mean value, which considers the range and distribution of all possible prices, is $4.17. There is nothing mathematically incorrect in using a price of $2.50 as an input for scenario analysis, but it is incorrect not to consider the effects of the range and distribution of prices and all other probabilistic input values in the cash flow analysis.

Even more important, the Monte Carlo simulation shows that there is a 90% probability that the NPV will be between $200,000 and $800,000; this is in contrast with the range of −$993,487 to $3,265,000 for the scenario approach. When the probabilistic distribution of NPV is considered, the project has positive NPVs through the entire range of probable values. Further, the median NPV of $532,180 in Table 5.3 is considerably higher than the most likely NPV of $5,692 suggested in Tables 5.1 and 5.2.

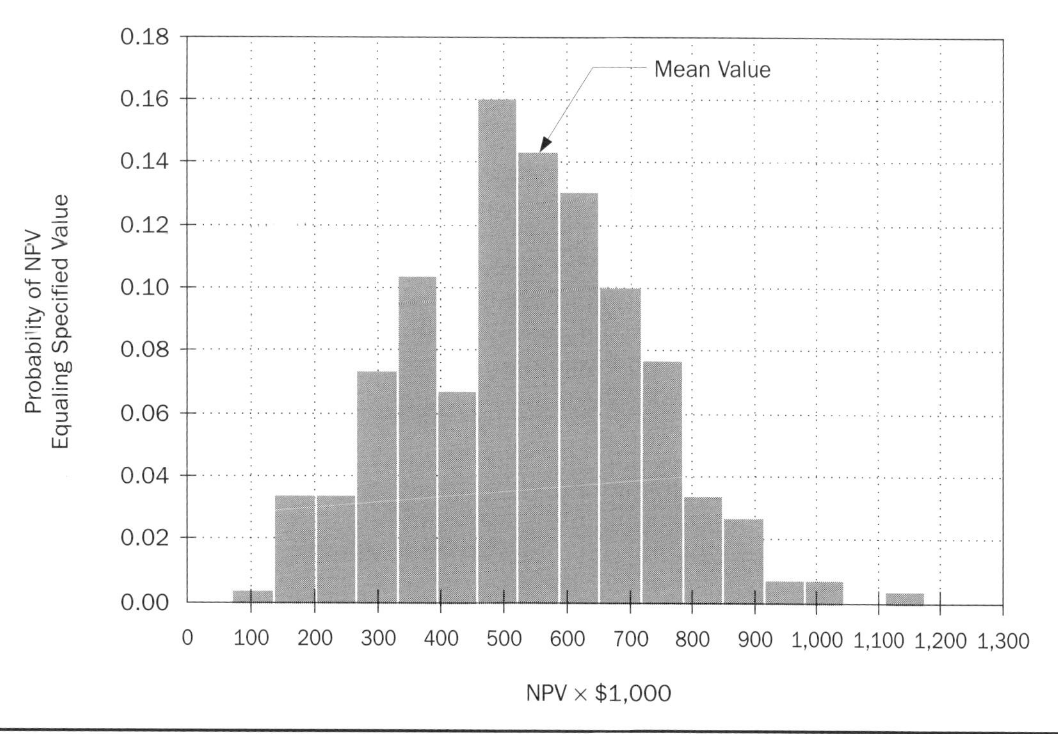

FIGURE 5.6 Distribution for NPV

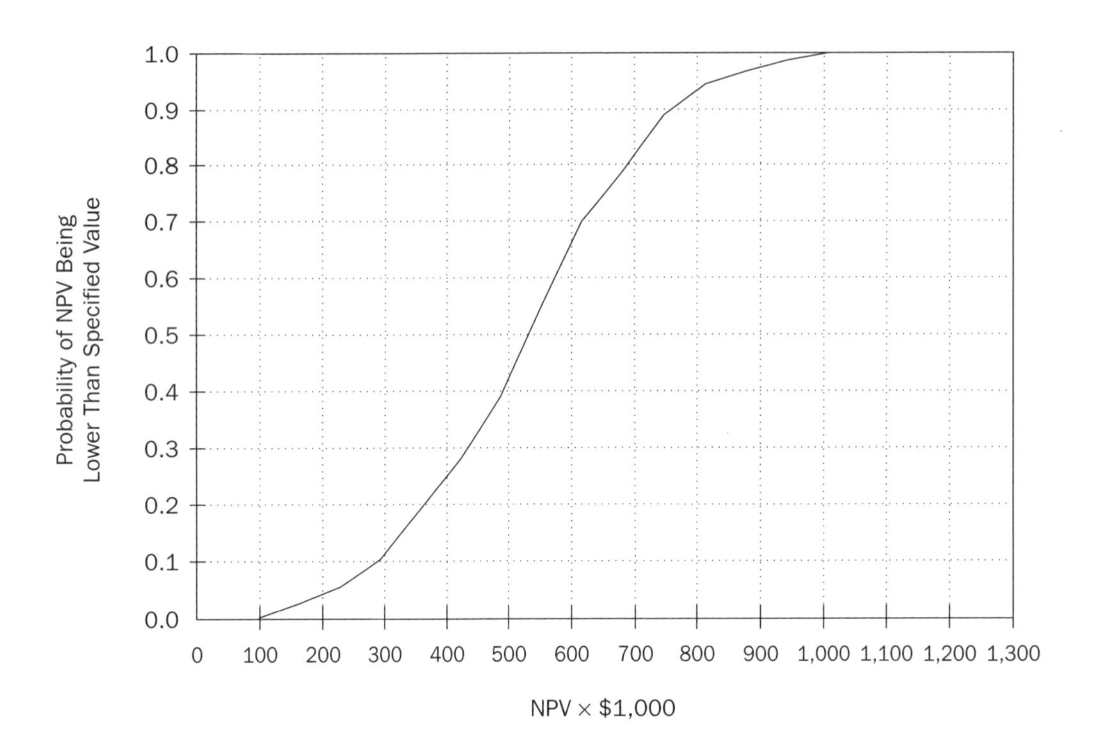

FIGURE 5.7 Cumulative distribution for NPV

TABLE 5.3 Mean values for probabilistic pro forma cash flow example

	Year			
	0	1	2	10
Product price, $		4.17	4.17	4.17
Quantity		60,000	60,000	60,000
Total revenues, $		250,000	250,000	250,000
Wage rate, $		14,333	14,333	14,333
Energy price, $		2.67	2.67	2.67
Materials cost, $		3.25	3.25	3.25
Number of workers		6	6	6
Units of energy		517	517	517
Units of materials		8,167	8,167	8,167
Total operating cost, $		109,142	109,142	109,142
Initial capital cost, $	333,333			
Cash flow, $	(333,333)	140,858	140,858	140,858

Median NPV for MARR of 10.00% = 532,180

Median payback = 3.0 years

The initial question can now be addressed: Is this a worthwhile project? Presented with the information obtained through the probabilistic simulation, a decision maker can easily see that this project is no loser even with the high discount rate of 10%. If a lower, more accurate risk-free discount rate were to be used, the project worth indicators would be even higher. However, the expected value of the project indicated in Table 5.3, even if a risk-free discount rate were to be used, is not the value an investor would actually pay for the project. That value cannot be directly obtained from the probabilistic simulation without consideration of the investor's attitude toward risk. The investor would incorporate risk into the probabilistic analysis to determine the project's certainty equivalence, which is the price the investor would actually pay. Probabilistic analysis simply describes the risk. The risk preference of the investor is then accounted for to interpret the risk and determine the price the investor would pay. The use of certainty equivalence is described later in this chapter and in Appendix C.

Monte Carlo Simulation Versus Scenario Analysis

There is much confusion about what the results of Monte Carlo simulation mean relative to those from scenario analysis. Assume the most likely case for scenario analysis is composed of the mean values for all individual inputs. The resulting NPV in this case is the expected value of the project. This expected value can now be directly compared to the expected value obtained from Monte Carlo simulation. For scenario analysis, a higher, risk-adjusted discount rate is used to determine the expected value; Monte Carlo simulation uses a lower, risk-free rate. This means that the expected value of the Monte Carlo simulation will be greater than that for the scenario analysis. Of course, value cannot be added to the project simply because of the analytical methodology used.

In the case of scenario analysis, the investor has indicated an opinion about the amount of risk inherent in the project through the choice of the risk-adjusted discount rate. The actual characteristics of risk may not be accurately identified in scenario analysis. In the case of Monte Carlo simulation, the investor has not indicated an opinion about the amount of risk because a risk-free discount rate is used. However, the actual characteristics of risk are revealed to the investor through the distribution of the NPVs of the many scenarios obtained through Monte Carlo simulation; it is then left to the investor to interpret the risk and decide what to pay for the investment. Therefore, the expected value from scenario analysis indicates the price the investor would be willing to pay considering the preconceived risks. The expected value from Monte Carlo simulation indicates the expected value of the project, which is higher than what the investor would pay because the risk preferences of the investor have not yet been considered. The investor must then assess the identified risks by using risk preference theory to determine the actual price to pay—which will be lower than the Monte Carlo–derived expected value.

If the discount rates are chosen correctly in both cases, both Monte Carlo simulation (after adjustment to account for the investor's risk preferences) and scenario analysis will indicate the same price that the investor would be willing to pay. The difference in the two methods is that risk is estimated on an ad hoc basis in scenario analysis and is determined through the identification and analysis of the risk characteristics of the project, as well as the investor's perception of risk, in Monte Carlo simulation.

Information obtained from a probabilistic simulation is superior to that obtained from a single-value or scenario analysis. In fact, both single-value analysis and scenario analysis can be extremely misleading even when proper care is taken in choosing the most likely, optimistic, and pessimistic values associated with comparable probabilities.

An example of idealized results from a probabilistic simulation is shown in Figure 5.8. This figure illustrates how a decision maker may use the NPV distribution to assess the possibility of loss as well as gains. The project has a 5% chance of losing $25 or more. However, it has a positive expected NPV and as good a chance of making over $100 as of losing over $25. Whether the project is desirable depends on specific circumstances, as interpreted by the decision maker. If the company cannot afford to lose $25 under any circumstances, it must reject the project. If the company can afford a $25 loss, though, it should accept the project. Therefore, the decision maker must consider multiple factors: the probability and amount of maximum loss, the probability and amount of maximum gain, the amount of the expected NPV, and the attitude toward risk.

Figure 5.9 shows how two competing investment opportunities can be compared on the basis of the NPV distributions obtained by probabilistic simulation. In this example, both projects have the same expected NPV. However, project A has a greater potential to make more and to lose more than project B. Again, this analysis involves multiple attributes that the decision

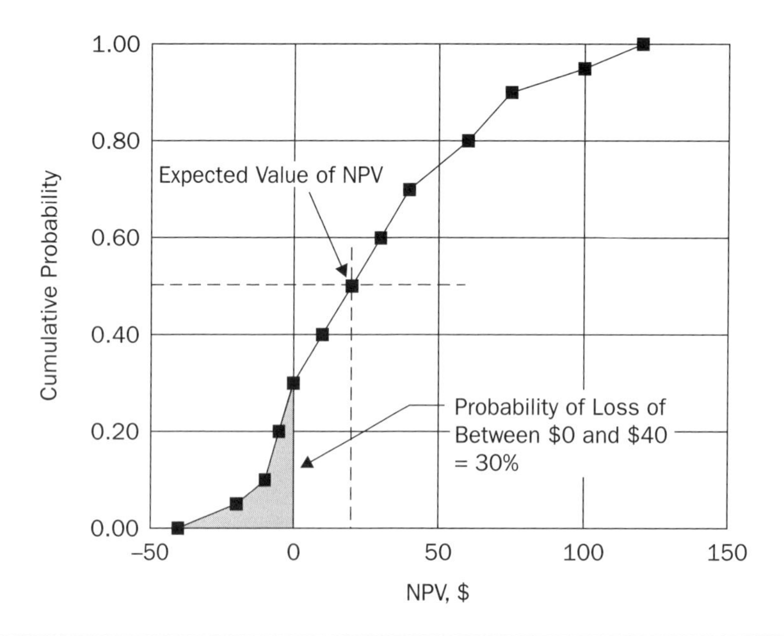

FIGURE 5.8 Using a cumulative NPV distribution to assess possibility of loss

maker must weigh. The NPV distributions themselves do not indicate which project should be undertaken.

Disadvantages and Constraints

Probabilistic analysis is superior to single-value or scenario analysis, but it is not used as widely by the mining industry as would be expected. To understand why this is so, the disadvantages and constraints of probabilistic analysis must be considered:

1. Conducting probabilistic analysis was cumbersome before add-on spreadsheet programs became widely available for computers. However, evaluators can now easily perform probabilistic analysis with relatively inexpensive software and appropriately outfitted personal computers.
2. Probability distributions of the input variables are often thought to be difficult to obtain. While this may be true in certain instances, an analyst can often simply discuss cost and sales estimates with engineers to obtain the needed distributions. In any case, analysts must identify appropriate probability distributions (high, medium, and low) for scenario analysis, and identifying these distributions is no easier for scenario analysis than for probabilistic analysis.
3. A difficulty in using probabilistic analysis is that it is necessary to identify the degree of correlation among the variable inputs. For example, product prices and energy input prices (such as the world oil price) may be correlated. In this case, it would be incorrect to treat both product price and energy input price as independent variables. This difficulty can be a valid

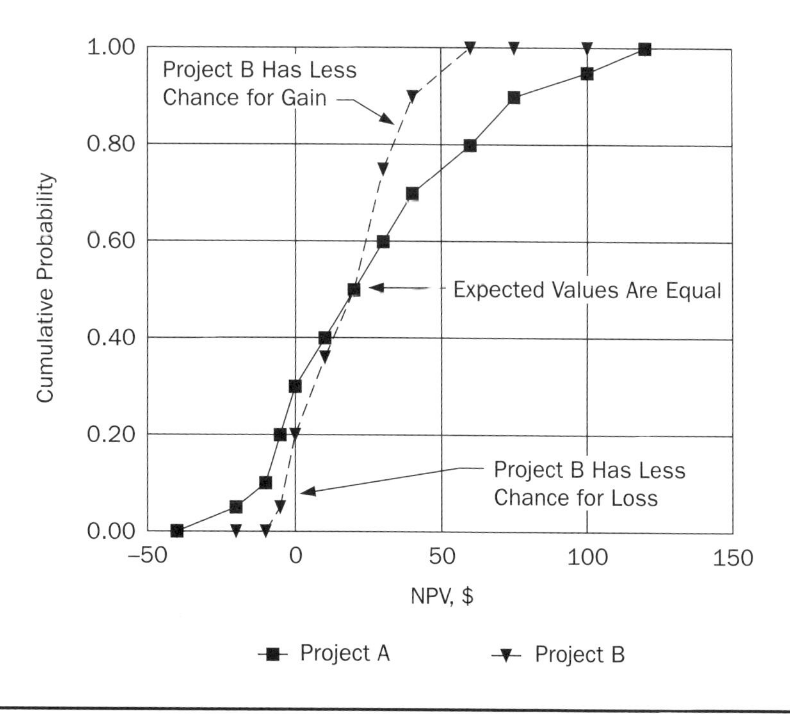

FIGURE 5.9 Probabilistic comparison of two projects

hurdle. However, probabilistic software programs have provisions to accommodate subjective estimates of correlation among the variables. The importance of correlation is case sensitive and can be tested with respect to the degree of importance. Correlation is equally significant for scenario analysis. In other words, a correlation problem is not a good reason not to use probabilistic analysis. The degree to which more precise estimates of correlations affect the determination of the NPV distribution is case specific and no other general conclusion can be made. Often only a few of the variables are correlated, and sometimes the correlation can be adequately determined by subjective estimation methods. It is possible to conduct a sensitivity analysis using probabilistic simulation to test the importance of correlation among variables. If the degree of correlation does not significantly affect the value of NPV, then the correlation is not important.

4. A more basic weakness of probabilistic analysis is associated with the static and inflexible nature of DCF analysis in general and involves the dependence of a variable on time. Many mineral prices do not generally fluctuate randomly from one year to another; instead they follow cyclical patterns that usually involve more than 1 year. Conventional probabilistic DCF analysis does not allow for a rational production response to changing prices as would be expected from a mine manager in real life. In this respect, the distribution for net present values will always be biased toward the low side (i.e., undervaluing a project). A simulation model that allows for production decisions in response to stochastic prices and costs would solve many of the deficiencies of conventional Monte Carlo simulation; for examples, see McKnight (1988) and Sarkarat (1996). In any case, problems remain with the choice of discount rate and the possibility of the discount rate changing over time.
5. It takes more time to conduct probabilistic analysis than scenario analysis, and analysts generally have many other projects requiring attention. However, the marginal amount of additional time required is small. When large projects are involved, the marginal benefit of using probabilistic analysis is much greater than the marginal cost of increased time requirements. On the other hand, probabilistic analysis of small, uncomplicated projects might require more time than could be justified.
6. Many evaluators and even more decision makers do not understand probabilistic evaluation and are more comfortable with scenario NPV analysis. This situation was certainly more prevalent 20 years ago; modern managers are commonly more familiar with advanced decision methods.
7. Probabilistic analysis does not give a single decision signal as does a simple NPV or IRR. This is particularly evident when two or more projects must be compared and ranked. Probabilistic analysis yields results for multiple attributes: amount of maximum loss, amount of maximum gain, and amount of expected gain. The NPV distributions themselves do not indicate which project should be chosen. The investor must know how to interpret the results. Paradoxically, it appears the main reason Monte Carlo simulation is not more widely used is also the methodology's major strength: the greater amount of useful information.

For more information on interpreting the results of Monte Carlo simulation or developing input data for the simulation, see Barnes (1980), Davis (1995a), Davis (1995b), Pindred (1995), Newendorp (1975), Gentry and O'Neil (1984), Stermole and Stermole (1993), and Torries (1996).

RISK ATTITUDES AND CERTAINTY EQUIVALENCE

An important characteristic of project evaluation is that different investors have different tolerances for financial risk. Determining these differences and acting upon them in practice are difficult. For example, an explorationist may be willing to fund high-risk exploration activities, while most banking institutions will not. These differences in risk taking reflect inherent differences in attitudes toward risk among different individuals and firms. These attitudes are not right or wrong but simply different for each organization for any number of reasons.

Probabilistic analysis is of little use if the investor does not correctly analyze and interpret the probabilities of success and failure, as well as the magnitudes of the payoffs. Efforts to avoid some of the pitfalls associated with the use of NPV and IRR in undertaking probabilistic analysis lead naturally to a discussion of a fundamental decision science model known as preference theory. This theory is an extension of the expected value concept in which the investor's attitudes and feelings about money are incorporated into a quantitative decision model. The result is a more realistic measure of value among competing projects characterized by risk and uncertainty.

The application of preference theory in determining project worth is perhaps not as well known as such evaluation criteria as NPV, IRR, and expected value, but the concept is not new. Daniel Bernoulli attempted to quantify an individual's feelings about money and risks in 1738, and in 1944 John von Neumann and Oskar Morgenstern developed a mathematical theory that addressed the subject (von Neumann and Morgenstern 1953). Preference theory concepts are based on some very fundamental and reasonable concepts about rational decision making that are well documented in the decision science literature (Savage 1954; Pratt 1964; Howard 1988) and apply particularly well to mineral project evaluation.

Preference theory is appealing in that it enables the investor to utilize a relatively consistent measure of valuation across a broad range of risky investments. The theory provides a practical way for investors to formulate a consistent policy that incorporates their attitudes about participating in risky projects. Under certain conditions, this risk attitude can be specified and is measured in the form of financial risk tolerance (RT). The RT value represents the sum of money such that the decision maker will just be indifferent about whether to accept an investment that has an equal chance of gaining that sum or losing half that sum. RT assessment and an example of its use are described in Appendix C.

Preference theory provides a means of mapping the investor's attitude about taking on financially risky projects in the form of a utility function. If evaluators know the investor's utility function, which measures the investor's preferences for uncertain outcomes, they can then compute a risk-adjusted valuation measure for any risky or uncertain investment. This valuation measure is known as the certainty equivalent; it is defined as the value, known with certainty, for which an investor would be indifferent about swapping that value in exchange for a risky project. For example, an investor may be indifferent between accepting (1) a project that has a 0.2 chance of gaining $20,000 and a 0.8 chance of losing $5,000 and (2) a sure opportunity gaining $8,000. In this case, the investor's certainty equivalent for the risky project is $8,000. Certainty equivalence accounts for the uncertainty in the project as well as the investor's appetite for risk and preferences for the consequences. Project comparisons using this approach are more meaningful than those using only expected value because expected value describes only the risk of the project and does not consider the investor's attitude toward risk. The decision criterion under the certainty equivalence approach is to accept the project(s) with the highest certainty equivalent(s).

The certainty equivalence valuation also provides guidance to the investor in terms of the value of diversification and risk sharing. Unlike expected value analysis, which is based on an "all or nothing" decision rule, certainty equivalence valuation aids the investor in selecting the appropriate level of participation in a project consistent with the investor's risk propensity. It provides a formal means to quantify the advantages of selling down or "spreading the risk." An example application of the preference theory concepts is described in Appendix C. For additional information on preference theory and its applications in the mineral industries, see Walls and Dyer (1992) and Newendorp (1975).

BAYESIAN DECISION MAKING

Bayesian decision making (also known as decision tree analysis) is another method to deal with the uncertainties that exist in forecasting cash flows over an extended period of time. This method recognizes that risk exists, that unexpected events do occur, and that a mine operator will react to these events. Conventional DCF analysis is a static method of evaluation because it involves only a single decision at the time of the evaluation. The project is either accepted or rejected under the specific assumptions included in the DCF model. Bayesian analysis is dynamic and is not confined to a single decision. Values are determined for each of the possible solutions identified in the analysis. The construction of a decision tree requires that the analyst is able to identify appropriate decision nodes and to assign probabilities to their likelihood of occurrence.

While decision trees enable an analyst to model the effects of combinations of probabilistic events, such as lower ore quantities, higher ore grade, and lower metal prices, the exercise can easily result in a large and very complex model. As a consequence, decision tree models are usually kept as simple as possible because of the practical limitations of computer capacity and time constraints. However, simple models run the risk of omitting important decision stages. Another problem is in the choice of appropriate risk-free discount rates for the decision nodes; risk is considered in the probabilistic decision analysis, but it decreases over time as unknowns are eliminated.

For more information on Bayesian analysis, see Hirschleifer (1961) and Ross-Watt and Mackenzie (1979). See King (1971) and Newendorp (1975) for specific applications of Bayesian analysis in petroleum exploration decisions and Gentry and O'Neil (1984) for examples related to mineral projects.

ACCOUNTING FOR POLITICAL RISK

Political risk is another area of concern. It covers a wide range of possible actions, potential costs, and remedies from the viewpoint of both the government and the investor. Excessive taxation, or government "take," is a major concern of project investors operating in many countries. There are many ways a government can obtain its take, including equity positions; exchange rate restrictions; limits on repatriation of profits; and the imposition of a wide variety of income, import, and export taxes, as well as royalties. Some governments collect portions of their take by charging higher fees for state-supplied transportation services, such as rail transport. Another important factor is the government's treatment of capital allowances such as depreciation and depletion. The government legislates these allowances to induce or restrict investment and to increase or decrease tax revenues.

The different ways of obtaining the take may have different economic effects on a mineral project, e.g., by restricting future mineral development, by encouraging rapid depletion of the project's resources, by creating a reduction in the employed labor force, or by increasing production costs and lowering international competitiveness.

Several ways to account for political risk in the evaluation analysis have been proposed. A number of political weighting schemes that result in a ranking of countries have been developed, with varying degrees of usefulness. In addition, corporations may increase the discount rate to account for the differences in political risk among projects in various countries (Humphreys 1996). For a summary of political risk analysis, see Chermak (1991).

There are a number of ways political risks can be reduced, but all methods to compensate for political risk have a cost. Insurance can be purchased to compensate for losses due to expropriation. Excessive government taxation or interference can sometimes be minimized by the involvement of such organizations as the World Bank or a number of international partners in the project. However, the best way for excessive government risk to be reduced is through the actions of the government itself. A government must be able to identify long-term and short-term economic rents available and to fully understand the consequences if the take exceeds these rents. A demonstrated history of reasonable treatment of mining projects is the best way for a government to reduce the political risk perceived by investors, as well as the attending costs of risk avoidance.

CHAPTER 6

Timing and Investment Options

Two important aspects of investing have not been addressed by any of the evaluation methods discussed so far: (1) being able to respond to fluctuating prices by postponing investments and (2) being able to exercise operational flexibility. Cyclical and fluctuating mineral prices are often considered an undesirable source of risk in mineral project investment, but they give investors opportunities to earn large rewards. This section addresses the cyclical behavior of mineral prices and introduces the concept of using option-pricing methods to evaluate mineral investment opportunities.

ACCOUNTING FOR CYCLICALITY

Mineral prices, especially those for metals, can exhibit a pronounced cyclical pattern. Since prices have been cyclical in the past, it is reasonable to believe that this pattern will continue in the future. Expected price fluctuations present particular difficulties in using conventional evaluation methods such as DCF analysis. One problem is that an evaluator may not know where current prices are in the cycle or when the next cycle will occur. DCF analysis requires forecasting prices, and the prices forecast during the first 10 years greatly affect NPV. If the estimate of the current point on the price cycle is incorrect, the evaluation analysis may be fatally flawed.

Another problem with conventional DCF analysis is that it makes no allowance for the option of closing an operation or changing operating methods during times of low prices. Ignoring this option in the DCF analysis may cause project values to be understated.

The results of having cyclical cash flows and the option to close operations during periods of losses are shown in Figures 6.1 and 6.2. Figure 6.1 shows a typical cyclical cash flow pattern over a 10-year cycle period. The option cash flow illustrates the cash flow when the operator exercises the option to close the operation during periods of losses, which occur in years 8, 9, and 10 in Figure 6.1. A comparison of the NPVs of the base and option cash flows generated when the

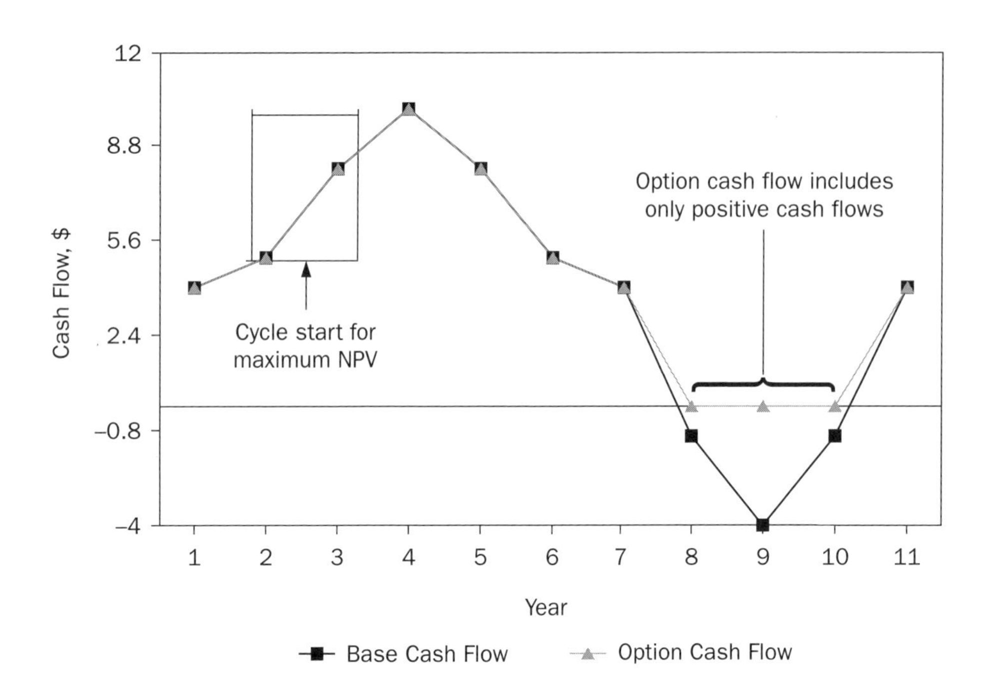

FIGURE 6.1 Cyclical cash flow pattern

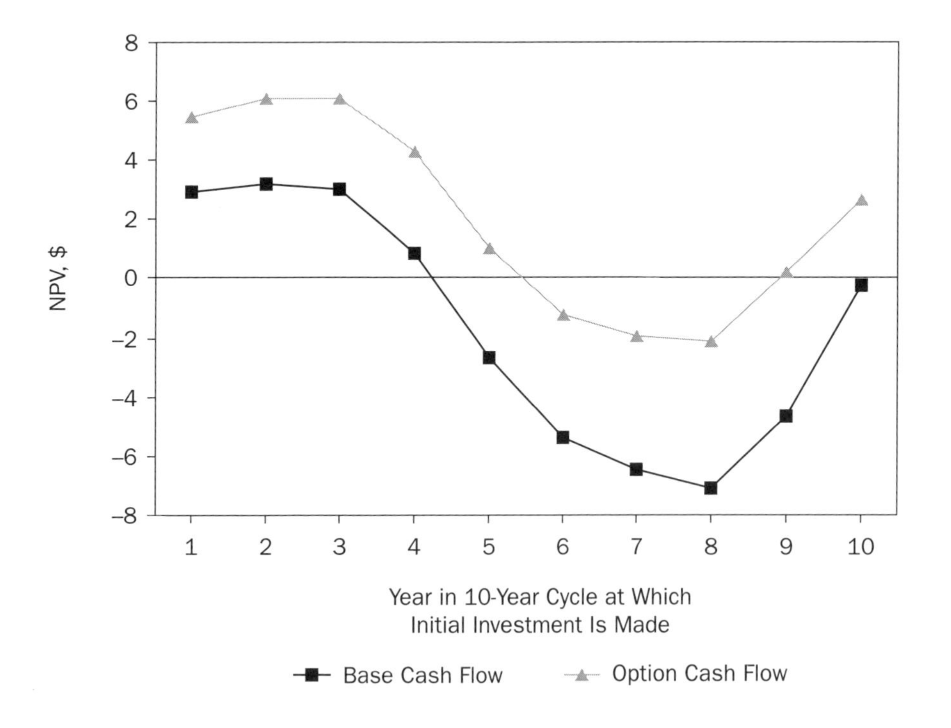

FIGURE 6.2 NPV and timing of initial investment in a cyclical price pattern

project is initiated in any phase of the 10-year cycle is shown in Figure 6.2. As the figure shows, NPV increases significantly when the option is exercised to shut down and avoid the negative yearly cash flows, regardless of when the project is initiated relative to the price cycle.

THE IMPORTANCE OF TIMING

No matter what method is used to evaluate a mineral project, there is no better way to ensure success than to make a correctly timed investment. Catching an upward surge in the price of the commodity soon after making an investment can cause inaccuracies in a pro forma discounted cash flow to become relatively unimportant. The reason, of course, is that the higher cash flows in the early years of the investment are the most significant in determining NPV and profitability of the project. The reverse is also true: A poorly timed investment almost always yields disastrous results.

In the base cash flow in Figure 6.2, NPV varies from $3.24 to –$7.09. This extreme variation is caused solely by starting the project at various stages in the cycle and clearly illustrates the importance of investment timing. Note also that the maximum NPV in the example is obtained by initiating the project not when the cycle is at its low zor peak, as is often thought, but rather when the cycle is well into the upswing. Investing at the trough of the price cycle involves obtaining relatively low, although increasing, prices early in the cash flow.

The exact effect timing has on NPV depends on the rate of increase and decrease in the cycle as well as the amplitude and frequency of the cycle. Increasing the discount rate amplifies the effects of the investment timing.

Since mineral prices can be cyclical, the method used to evaluate a mineral project is perhaps of less importance than the ability to correctly recognize troughs and peaks in the price cycle. Timing alone does not represent the single decision criterion to be used in the evaluation of an investment, but it is one of several an investor should consider. For additional reading on feasibility tests and business cycles, see Kaufmann (1983).

OPTION PRICING

There have been attempts to develop evaluation techniques that recognize the existence and effects of price cycles and a firm's ability to respond to unexpected changes. One such method is known as option pricing. An option is the right, but not the obligation, to buy or sell a commodity at some time in the future. Since this right has a value, options have a price. Therefore, purchasing an option for the promise of future delivery of a mineral commodity is similar to purchasing a mineral project with the intention of selling the mineral at some time in the future. The greater the expected price fluctuation, the greater the value of the option (and the less accurate DCF analysis is likely to be). With option pricing, the analyst's task is to determine the value of a bundle of options that represent the same commodities, risks, and expected price fluctuations as the mining project in question and then to use the value of the combined options as the value of the mining project.

Option pricing is also useful in determining the value of postponing decisions concerning irreversible investments with uncertain payouts that may be realized far in the future. For example, electricity producers are currently evaluating how to meet uncertain future demand for power. While a number of technical alternatives are available to the electric power industry, consider only two to illustrate this use of option valuation: (1) the construction of several small low-capital-cost but high-operating-cost units, such as gas turbines, or (2) one large high-capital-cost but low-operating-cost unit, such as a hydroelectric project. Since the electric utility does not know future demand with certainty, constructing the large hydro facility may be either a financial success or failure. There is a cost associated with making the decision to construct the hydro project because the investment is irreversible. As long as the option to construct the hydro project still exists, choosing the other option (the gas turbines) allows the possible cost of making an ill-advised investment to be avoided. The problem, then, is how to incorporate the value of postponing the decision to invest in the large project. Conventional cash flow analysis completely fails to capture the value of the option to make or not to make the investment. However, determining the value of this option is exactly the purpose of option price analysis. Electric utilities are currently using option pricing to evaluate their expansion options.

Although option pricing as a method of project valuation is not widely accepted at present, its popularity is growing rapidly. The methodology is more complicated than DCF analysis, and the concept and evaluation results are less easily grasped. In addition, option pricing is based on knowing the future expected variability of prices. Here again the problem of whether it is valid to extrapolate past trends into the future arises.

Option-pricing methodology may not be appropriate for all investment situations. However, combining option-pricing methods with DCF analysis, Monte Carlo simulation, decision trees, and mathematical programming promises to become extremely useful for project evaluation.

There are numerous sources of information on using option-pricing methods to evaluate natural resource investments. The basic method of option pricing was developed by Black and Scholes (1973). The use of option pricing to evaluate natural resource investments is discussed and well illustrated in Brennan and Schwartz (1985). McKnight (1988) compares Monte Carlo simulation that includes operational decisions to conventional DCF analysis and to option valuation. Additional information on the use of options pricing and contingent claims analysis is given in Guzman (1991), Lehman (1989), Paddock et al. (1988), Siegel et al. (1985), Palm et al. (1986), Mason and Merton (1985), and Cavender (1992), Trigeorgis (1996), and Davis (1996). Also see Appendix B of this text.

CHAPTER 7

Overviews of Other Evaluation Methods

Numerous other types of evaluation methods exist to determine the effects or feasibility of mining activities. The more widely used methods are described in this chapter.

MEASURING ENVIRONMENTAL BENEFITS AND COSTS

Society and governments are increasingly requiring the inclusion of environmental effects in the monetary valuation process of private industrial projects, as well as social projects. The problem faced by all analysts is that it is difficult to value environmental attributes, which often are not traded on the market, such as clean air, the preservation of a beautiful view, or the preservation of global biodiversity. Four classes of models are now being used to value these types of goods: market-derived prices, hedonic analysis, contingent valuation (CV) methodologies, and travel models. Both industry and governments need to know the details of how these methods of analysis work because each methodology has a set of strengths and weaknesses. As with any other valuation tools, these methods can easily be misused or the results misinterpreted.

Market prices are indications of the willingness to pay (WTP) on the part of the buyer and willingness to accept (WTA) on the part of the seller; they are the easiest and most direct valuation techniques that can be used to evaluate environmental costs. However, this methodology can be used only for those environmental goods that are traded on the market, such as tradable air pollution permits. Even if market prices exist, they will not reflect the true environmental cost unless there are no externalities and no underutilized resources.

Hedonic pricing of environmental factors involves the use of prices of other goods affected by these factors as indicators of the factors' values. For example, homes close to a noisy airport or near a bad-smelling factory have values less than those of similar homes that are not. On the other hand, homes that have great ocean views command greater prices than for similar homes that do not. While there is no direct market price for noise, odor, or views, their values can be determined by observing the differential values in houses. The hedonic pricing approach, which identifies the WTP for pleasure or the WTA for not having pleasure, was initially developed by Ridker (1967). This methodology is widely used by appraisers and the real estate industry to determine the values of homes and commercial locations. It requires much care in determining and analyzing the hedonic values to avoid statistically incorrect and highly misleading results.

The CV methodology involves questioning a statistically significant portion of the population to obtain the public's views on the subjective values of WTP or WTA, either through direct interviews or by questionnaires. Interview or questionnaire design is a critically important component of CV methodology. CV is particularly well suited to address items not traded on the market, such as environmentally or socially oriented items. CV can also be used to determine the value of nonuse of items, such as the worth to an individual of not being exposed to safety risks or not having a shortened life span. The purpose of determining the values of nonmarket goods through CV is to incorporate them in conventional DCF analysis or other types of valuations. For additional information on the uses of CV, see Mitchell and Carson (1989) and Hanneman (1994). For skeptical views of the usefulness of CV, see Arrow et al. (1993) and Diamond and Hausman (1994).

A travel model considers the amount people will pay to travel (1) to enjoy an environmental good, such as a national park or a fishing stream, or (2) to avoid an environmental bad, such as urban smog and traffic. For example, a travel model determines the value of a good trout stream by identifying the cost fishermen would be willing to pay to travel to fish in the stream. Data for travel models are always derived from costly and time-consuming surveys. The design of the survey is critical to the successful determination of value.

These four methodologies can be used to estimate environmental costs if properly conducted. However, such evaluations take time and resources. Poorly conducted evaluations give results that appear suitable but are likely to be erroneous and misleading, which is worse than having no results at all. For additional information on the valuation of environmental factors, see textbooks on environmental economics, such as Lesser et al. (1997) and Tietenberg (1998).

INPUT/OUTPUT ANALYSIS

Input/output analysis (IO) is a method of accounting that can be used to determine how changes in one sector of the economy affect the other sectors and the national economy as a whole. Thus, IO makes it possible to determine the effects a new mineral project might have on the economy of a nation.

IO reports both primary and secondary effects of a new or changed sector of an economy, thereby providing a true picture of the total effects of the new economic activity. For example, a new mine will directly employ a certain number of people but will also create employment indirectly for those providing goods and services to the new mine. Therefore, the total amount of employment gained from the opening of a new mine is often 1.5 to 2.5 times the amount of labor directly employed by the mine. This multiple, called the employment multiplier, changes according to the type of mine and related processing. From a country's viewpoint, higher employment multiples are usually preferred because of the country's need to reduce unemployment. Income multiples, which measure the increase in income a new project is likely to have on a region can also be determined.

IO also provides a means to tell how a new mining venture will affect other sectors of the national economy. For example, a new mine might buy coal-generated power to operate its equipment and mills. IO could be used to determine the economic effect from the expected increase in activity in power generation and coal mining. If the coal-mining industry of a country were particularly depressed, this information would be of great interest to government decision makers.

IO is a powerful tool, but like all others it has a number of drawbacks. The use of IO requires that a table of coefficients showing the relationship among the industries in a country be available. Many countries do not have such tables, and constructing new tables is sometimes difficult and expensive. In addition, these tables become obsolete if not updated periodically. Obtaining reliable coefficients for a country or region is difficult even under the best of circumstances. IO is typically used as a static analysis, although modifications using dynamic IO have overcome some of the associated problems. For a further discussion of IO and its use in mineral project selection, see Torries et al. (1988).

ECONOMETRIC MODELING

Any economy can be represented by a system of equations that form what is called an econometric model. The equations are obtained by a statistical examination of past relationships among supply, demand, and prices of goods and services in the country. These equations and the resulting econometric model can be used to determine how changes in portions of the economy might affect other sectors and the economy as a whole. As a result, the effects a new mine or mineral activity would have on an economy can be modeled. In addition, the model can be used to tell what effects changes in taxes or prices might have on the demand or supply of commodities, as well as the resulting secondary effects, such as changes in national employment and income.

Private industry does not generally use econometric models in the same manner as it uses DCF analysis to evaluate individual mineral deposits. Econometric models are more commonly used by governments to determine the effects a new activity might have on the economy. As with IO analysis, obtaining accurate coefficients for the equations is a major task.

MATHEMATICAL PROGRAMMING METHODS

A problem may exist in determining which mix of projects is to be chosen given budgetary and operating constraints. This problem can be solved by the use of integer programming, which is a branch of linear programming. In this application, linear programming is a mathematical method used to find the set of projects that yield the greatest total NPV given budgetary and operating constraints. Integer programming is required to prevent the solution from including fractional projects, which is not possible.

Mathematical programming can also be used to determine the least-cost (or maximum-profit) mix of projects to be undertaken given the nature of the individual projects and the relevant constraints. A good example would be to choose the optimal mix of power generation methods given the costs and capacity for each technology and the overall required capacity and constraints.

A project often has, more than one desired goal, e.g., profit maximization and pollution minimization. This type of problem can be solved using multiobjective or multigoal programming (Blair 1979). While multigoal programming appears to simulate real-life conditions of having multiple (and sometimes conflicting) goals, determining the weight or importance of one goal over another is necessary. This process is subjective and is subject to criticism as a result. Much of the criticism is answered if sensitivity analysis indicates that the value of the project is insensitive to variations in the weights over a reasonable range of values.

FINANCIAL ANALYSIS AND DEBT

Financial analysis consists of finding the least-cost method of obtaining funds to finance the construction and operation of a project. Financial analysis is conducted after the economic analysis indicates that the project is economically sound and should be undertaken. It is commonly assumed that if a project is not feasible in the absence of debt financing, it will not be feasible under any financial arrangement. However, this assumption is not always accurate; a favorable and highly leveraged financial arrangement can indeed influence the economics of a project.

It is worth noting here the differences between equity and debt. Equity investments are made by risk-accepting investors, and the compensation for doing so under the risks inherent in the project comes from any profit the project may make. There is no guarantee that there will ever be a return on an equity investment. Debt, regardless of form, consists of a legal contractual agreement by which the lender agrees to lend the funds and the recipient, or project participant, agrees to repay the loan plus interest. Debt can take the form of a number of possible instruments, including bonds, preferred stock, commodity and equipment loans, and loans made by lending institutions.

Contractual agreements for loans may be complicated and can contain numerous covenants that are designed to protect the lender's interests but usually restrict the borrower's actions until the loan is repaid. Common provisions include a series of ratios, such as a maximum debt-to-equity ratio or reserve-to-production ratio. Ratios and ratio analysis are important issues in negotiating mineral finance. For additional information, see Schreiber (1985) and Castle (1985). The dynamic nature of capital budgeting presents problems with respect to the timing of expenditures. A summary of dynamic investment criteria is given in Thompson and Thuesen (1987).

ARE SOPHISTICATED METHODS WORTHWHILE?

The existence of so many evaluation methodologies, the many traps in conducting the analyses, and the innumerable problems with the data are sufficient to confuse even experienced evaluation practitioners. While increasingly complex evaluation methods are being developed to compensate for these difficulties, one may well wonder whether complex models are really worthwhile given all the imperfections in the data. In any case, it is not entirely clear which of the methodologies may be best.

It is difficult to prove that the more advanced methods give better results. Computer spreadsheets and the more sophisticated evaluation methods have become available on a widespread basis only recently.

One way to determine the worthiness of the project evaluation process is to conduct a postproject audit and to compare the original rate of return with the realized rate of return. While the concept is clear, conducting actual comparisons is difficult, as shown by Pohl and Mihaljek (1989), who analyzed the results of 1,000 World Bank projects completed between 1968 and 1980. Results of that study were mixed for a number of institutional, procedural, and market-related reasons. Although there was no clear consensus, Pohl and Mihaljek (1989) concluded (1) that benefit:cost and rate-of-return analyses are most useful in screening out large, capital-intensive projects with potentially low rates of return and (2) that the inclusion of shadow pricing in the analysis is not worth the trouble. Little and Mirrlees (1990) commented on the results of Pohl and Mihaljek and concluded that, in spite of the diminished value of shadow price evaluation, the effort is worthwhile. They also argued that project evaluation has merit even in the face of severe uncertainties, and they presented an analysis to determine the value of project appraisal.

Humphreys (1996) offered another view of the usefulness of project and risk evaluation: that corporations often use conventional DCF analysis and a predetermined hurdle rate as a screening device on large investment opportunities to make "go/no go" decisions. These types of investment decisions do not require a specific identification of the value the market would place on an asset. This approach avoids many of the difficulties associated with accurate price forecasts. If the project appears to be highly profitable under a variety of plausible scenarios, then the investment is made. This is truly a case of making the investment on the basis of NPV > 0 or IRR > MARR. Specific values are needed when a corporation is considering merger, acquisition, or divestiture.

Humphreys also pointed out that the value of flexibility of operation is sometimes not a consideration for large operations because it is often not operationally feasible to shut down and restart in concert with the market. He suggested that large mining operations are designed to run at full capacity and that, under conditions of declining prices, a corporation may as well choose to make additional investments to reduce operating costs rather than to reduce output. On the other hand, some large companies do extend planned summer vacation shutdowns during periods of low prices.

On balance, it appears that more sophisticated models do give better evaluation results than simplistic models. Complex models do give more accurate estimates of the actual worth of investment opportunities under certain conditions. More important, complex models also allow a better understanding of the interaction of the complex factors that determine project feasibility and give investors more information on which to base decisions. Furthermore, the development of more complex evaluation methodologies leads the way to future improvements in the accuracy of evaluation results. For discussion on the usefulness of complex evaluation procedures, see Pike (1989) and Humphreys (1996).

CHAPTER 8

An Evaluation Guide

This text has discussed the goals of project evaluation, evaluation methods, and the types of responses by investors in interpreting evaluation results. What remains is to describe a generalized series of steps required to completely evaluate a project, as well as the potential errors that analysts may make at each crucial stage of the investigation.

STEPS IN PROJECT EVALUATION

Although the specifics of a mineral project dictate the specific evaluation method and particular steps that must be taken, the following text outlines a generalized set of steps for conducting a conventional DCF investment evaluation plus any suitable, more sophisticated analytical extensions. It is offered as a suggestion only; "cookbook" approaches to project evaluation do not exist. What is certain is that there is no single best route to project evaluation and that there will be numerous cycles of the evaluation as additional information is obtained. It is important to remember that project evaluation is a process rather than an event.

1. Determine all the organization's goals and its perception of acceptable and unacceptable risks. This step may be revisited a number of times as more information is obtained on the project in question. This vital step is one that is frequently slighted. No one can conduct a meaningful evaluation without a clear knowledge of what is desired.
2. Use the identified goals to identify the decision criteria, such as the minimum acceptable risk-free rate of return and the acceptable levels of risk based on the potential for increased benefits. While profit maximization is commonly used as a decision criterion, there may be other criteria that are also important. For example, a government might be interested in increasing employment or increasing foreign exchange earnings.
3. Identify all feasible alternatives, including doing nothing. Select those alternatives that are obviously more desirable for additional analysis.

4. Obtain data from existing geologic and technologic evaluations to obtain estimated grades, quantities, operating and capital costs, and prices for the specific project(s) in question. These data inputs represent the initial attempt at data collection and are more useful for identifying data collection problems than for obtaining actual evaluation results. It is helpful if estimates can be made of the expected variation in the data inputs. A preliminary evaluation using the annuity approach (see Appendix A) may give gross indications about a project's feasibility.
5. Identify and forecast all relevant variables, such as prices and costs, on a constant dollar basis. Start with rough estimates where necessary and refine them as the evaluation proceeds. Identify and quantify all relevant costs, including pollution and future reclamation and shutdown costs. On the production side, economies of scale–lower costs through increased capacity and production–are often important. On the demand side, markets and prices are important in that increased production must be balanced against increased supply and lower prices. The real problem in investigating demand factors is that analysts must generally use historical data for analysis of future events. The past may be the key to the present in the field of geology, but this may not be true in the field of economics and project evaluation. It will probably be necessary to use past cyclical price patterns and forecast levels of future economic activity and production costs to estimate future prices.
6. Construct a constant dollar pro forma cash flow with known taxes and royalties, and determine NPV, IRR, and payout. Although the real world has inflation, including inflation in the early cash flow analysis adds only unnecessary complications at this stage of the investigation. A discount rate must be chosen to determine NPV. A constant dollar hurdle rate of 6–8% might be appropriate, with further adjustments for risk. The risk-adjusted discount rates must be the same for all similar projects at similar levels of development. Different types of projects may require different levels of risk adjustment to the discount rate.
7. *Do a "reality check."* Do the results of the initial cash flow analysis make sense? How do these results compare with the initial analysis conducted in step 4? The evaluation process is complicated, so it is easy to make errors. It is always useful to use more than one method to evaluate a complicated project and to compare the results of the two methods. If the reality check fails, search for and correct errors in logic or accounting or improve the data; then repeat the process. This is an important but often neglected step in the evaluation process. It is wise to perform this reality check after the completion of each major portion of the evaluation.
8. Conduct sensitivity analysis to identify critical variables and possible floor and ceiling values of the project. Improve the quality of data for the critical variables, and redo the cash flow and sensitivity analysis if there are significant data changes. Test the effects of the discount rate in the sensitivity analysis. Conduct scenario analysis to test the effects of changed assumptions, but keep in mind the pitfalls. Consider expanding the scenario analysis into probabilistic analysis at this stage.

9. Write an interim project feasibility report suitable for the intended audience. This report should identify the basic key factors of why the project should or should not be further evaluated and what should be stressed in future investigations. It is important that a series of written reports be prepared during the project evaluation process. These reports formalize the process of forming and recording opinions and conclusions. This is particularly important for conducting scenario analysis in which numerous assumptions are changed and many cash flows are generated. It is easy to get lost in the many computer runs that often constitute a scenario analysis.
10. Determine if another method of evaluation, such as Monte Carlo simulation, decision trees, or option-pricing analysis, is appropriate to provide data on which to base a decision. This is particularly important when the investor has multiple goals that reflect a tolerance for risk or when the initial conventional DCF analysis shows a marginally unacceptable or marginally acceptable value for the project. If the analysis indicates the possibility of cyclical prices, be sure to consider the consequences of shutting down the project or changing mine methods or ore grades to control costs during forecast times of low prices and negative cash flows. This may entail simulation analysis.
11. In the case of a government evaluation of a social project, another analysis may be required to identify all relevant social benefits and costs. A government may wish to include input/output analysis, econometric modeling, or shadow prices.
12. Construct appropriate constant dollar balance and income statements to determine the effect the project would have on the organization as a whole. In the case of a social project, this is the equivalent of input/output analysis or econometric modeling.
13. It may be desirable at this point to forecast future inflation and exchange rates and construct a current dollar cash flow showing current dollar NPV, IRR, and payout. Because inflation reduces the real value of fixed deductions, such as depreciation and cost-based depletion, taxes will be greater and NPV and IRR will be less than in the constant dollar case. Current dollar analysis is also needed to show how inflation rates that vary over time or among the input or output factors change yearly cash flow values.
14. Make conclusions about the worth of the project from the viewpoint of the investor or organization. Make sure the analysis includes an indication of what might go wrong as well as right, an identification of the probability of these events happening, and the likely consequences of the events. Also make sure risk is treated in a manner consistent with other projects that the investor or organization is considering. Compare the economics of this project with other projects being investigated.
15. Make another report that is suitable for the intended audience and that discusses the overall feasibility of the project. This report combines the evaluation data with the goals of the investor and uses decision analysis to make the appropriate recommendation.
16. If the project appears feasible and if the project is large relative to the assets of the investor, it is usually necessary to conduct an analysis of debt financing. This is done by expanding the pro forma cash flow analysis to

include provisions for debt and debt interest rates. Determine appropriate finance options given the constraints, ratios, and covenants likely to be demanded by the lender. (Note that quoted bank rates are always current dollar rates and must be adjusted for inflation if they are to be used in a constant dollar analysis.)

17. If the project still appears feasible, determine the project worth from the viewpoint of other project participants, whether they are lenders, governments, or private corporations. Identify the distribution of economic rents among these participants. The purpose of this analysis is to obtain bargaining power for use in the negotiations that are likely to follow. This phase of the evaluation process is often neglected but can yield large benefits in terms of either increased capture of rents or decreased time of negotiation.
18. Reconcile differences in estimates of project worth and risks among the project participants and reach conclusions.
19. Draw conclusions about the project and write the final project evaluation report. Again, be sure to include an analysis of what might go wrong as well as right. Judge the risks in light of the risks experienced in other projects the organization is considering and the organization's attitude toward risk.
20. After the project is accepted and completed, conduct an audit of the project and the evaluation process to determine how the evaluation compared with actual experience. Although a subsequent audit of a completed investment is not really part of the evaluation process itself, it will provide additional information for future use.

POTENTIAL ERRORS IN EVALUATING PROJECTS AND RISKS

There is ample room in the evaluation process for errors caused by inappropriate procedures or an analyst's inability to foresee the future. The effects of individual errors can be great or small, depending on the magnitude of the errors, whether or not the errors are interdependent, and whether the errors are compounding or offsetting in nature. The expected combined effect of expected errors (i.e., forecasting errors) can be handled through the use of a risk-adjusted discount rate in the case of DCF analysis or by the use of other analytic methods such as decision trees. Risk cannot be reduced through analysis. However, analysis can highlight problems so that decision makers can take appropriate action to reduce risk, such as making long-term sales contracts, utilizing commodity and currency hedging techniques, or buying insurance.

Excessive Optimism

Individuals propose projects because they regard the undertakings to be worthwhile. However, there are often advocates or opponents who view a project from different perspectives and can sometimes point out important flaws. It is easy to get "caught up" in a project and become less objective than desired, particularly in the case of an individual or group of individuals responsible for the initial find or when the project has been under evaluation for a number of years and has required monumental efforts on the part of the evaluators. Objectivity in assessing both potential gains and losses is difficult to maintain as the evaluation process evolves over time, but it is nonetheless essential.

Current and Constant Dollar Errors

All items in a cash flow must be identified as being in constant or current dollars. Similarly, constant or current discount rates must be used consistently to correctly determine the value of the cash flow. The most frequent error made in DCF analysis is to use an inflated discount rate to evaluate a constant dollar cash flow.

Examples of this type of error are readily available in tax assessors' handbooks for numerous states. These handbooks recommend using the income approach to property valuation, whereby the forecast yearly income from a property is discounted using the weighted average cost of capital as the discount rate. Since the forecast yearly income usually consists of extending the historical level of income over a period of years, the forecast cash flows are in constant dollars. However, the discount rate is in current dollars. The fact that this inconsistent approach is common practice does not make it any less incorrect.

Another frequent error is to compare a constant dollar IRR with current dollar interest rates. For example, a series of equal cash flows over a period of years is often used as a basis for evaluating an investment. These equal cash flows and the calculated IRR are all in constant dollars. However, it is common for this constant dollar IRR to be compared with a current dollar MARR, such as a bond rate, to see if the project is economically favorable.

Geologic and Technologic Evaluation Errors

Errors in mineral project evaluation can easily be made during the geologic and technological evaluations, when reserve and technological risks are identified. The risks produced as a result of such errors influence operating and capital cost forecasts, project completion forecasts, and production forecasts of mineral quantities. A number of these factors can be interdependent, such as ore grade, amount of product produced, and operating cost. The geologic evaluation is particularly important because ore deposition and characteristics form the basis of all the steps that follow; these are the factors over which companies have absolutely no control. It is always possible to change mining or milling methods if errors are discovered in the technologic evaluation (at some capital cost, of course), but ore deposition patterns or grade cannot be changed in light of mistakes in the geologic evaluation. The best way to handle geologic risk is to include sufficient safety margins in ore grade and quantity forecasts.

Economic Evaluation Errors

Major errors in the evaluation of any large project are almost always introduced during an economic evaluation. Analysts expect to make mistakes in all forecasts, including those for product prices, environmental standards, and technological advances. In addition, errors concerning ore and product quality are common because of changing technology in ore- and material-processing technologies and product markets. Evaluators are often forced to use historical trends to make forecasts of the future, and this tactic often yields incorrect results.

Market risk, which includes the uncertainties of product prices and quantities, is a major concern in any project evaluation. It can be reduced through long-term contracts, by having customers as equity partners in the project, and by using hedging techniques and options. However, this concern with market risk may be misplaced. Perhaps, as suggested by Guzman (1991), companies should position themselves in a manner to take advantage of the expected changes in market prices, rather than trying to avoid the problem or assuming it does not exist.

Errors in the Choice of Discount Rates or Risk Evaluation

The proper choice of discount rate and the correct evaluation of risks, regardless of their sources, constitute major sources of errors. The perception of risk in a given circumstance is largely dependent upon the individual attributes of the investor, whether the investor is an individual, a firm, or government. For example, three individual investors with different propensities for risk would view the feasibility of a risky project differently.

Seldom do two investors view a single project in exactly the same manner. Furthermore, the perception of risk may change over time as a project develops. What is perceived to be risky at the start of a project, such as geologic data, may not be perceived the same way at some later date in the project's life. Risk perception may also differ greatly among individual investors depending on the amounts of potential losses and gains a project offers relative to the investors' wealth. To minimize the cost of risk assumptions, it is important that the varied risks are assumed by those project participants who are in the best position to handle–and reduce–them.

Errors that can result from choosing risk-adjusted or riskless discount rates depend on the type of evaluation and data. For example, NPV evaluations should, in general, use risk-adjusted data. For probabilistic analysis, on the other hand, a large portion of the risk is supposed to be accounted for in the probabilistic nature of the input data, and this portion of the risk must be removed from the discount rates.

Inconsistency in Risk Evaluation

It is well recognized that analysts must include risk in the evaluation of any project. It is not so well recognized that they must determine these risks in a manner consistent with that for competing projects. Otherwise, the project worths will not be derived in a consistent manner and cannot be accurately compared with each other. Evaluators must be careful in assessing risks across multiple projects because not all projects are similar in nature or at similar stages of development. For example, two gold deposits in the same geologic province of a country may be in different stages of development. The less developed one will pose a greater risk than the other, and the relative discount rates should reflect this relationship. The need for determining comparable risk adjustments for multiple projects is easy to understand, but the process can be very difficult in practice. For additional information on the perception of risk among projects, see Walls and Dyer (1992).

EVALUATION OF LARGE PROJECTS

The methodologies used and amount of effort expended to evaluate a project often depend on the size of the project. Large investment opportunities may have more potential to result in financial success or ruin than a single small opportunity and are generally worth additional evaluation effort. However, deals between major corporations, for example, are frequently the result of negotiations between senior executives who rely on experience and judgment to determine values in relatively short time periods because of external time constraints. This type of situation simply highlights the fact that human decision makers—not evaluation methods—make investment decisions.

Another characteristic of larger project investment opportunities is that they will not succeed without an advocate. Competition for funds is keen, and a strong and influential advocate is needed to obtain the resources for gathering the information required for the evaluation and to get the proper attention of senior decision makers. While it is often thought that good deals simply exist, waiting to be discovered, the truth is that deals are more often forged. Large favorable investment opportunities are typically brought about by the efforts of an advocate, such as a corporate officer. Evaluation methods may even be used to forge a deal rather than simply evaluate it.

Since an advocate is important in the successful outcome of a larger investment involving a corporation, government, or institution, it must be remembered that that individual may have a personal or other type of agenda that favors the investment opportunity. For example, a chairperson may have a wish to become the head of a billion dollar corporation and may let this desire influence acquisition activity. This is not to say that the resulting acquisitions are right or wrong, but it does suggest one reason financial evaluation cannot always explain actual investments.

It is worth noting that analysts can spend considerable portions of their professional lifetime in the evaluation of a single large project. Under these conditions, projects take on a life of their own as the evaluation proceeds. The natural human inclination in such case is for evaluators to encourage the successful development of the project. It is sometimes very difficult to completely reject a prospect that has been in the evaluation stage for many years. Whether this is good or bad depends on the circumstances. It is clearly undesirable to continue the investigation of an inferior investment opportunity. On the other hand, abandoning a project too soon is also undesirable. Many profitable investment opportunities have been forged over time through the tenacious efforts of the project advocates and analysts. It is not possible to suggest specific solutions for these situations because of the many unique and project-specific factors that are usually involved.

Postinvestment audits are useful tools to determine why some large projects succeed while others fail. Failure or success can occur for unanticipated or unknowable reasons, which may make the value of evaluation seem questionable. Postinvestment audits are costly, however, and do not give an immediate return on investment. Consequently, corporations or governments do not usually make the required effort.

CONCLUSIONS

Many methods of investment evaluation are commonly used. All have faults, and none are entirely adequate in all cases. Merit measures based on worth, as well as those based on rate of return, all provide useful information to a decision maker. However, evaluators sometimes forget that the analysis simply provides information—it does not and cannot choose one project over the other. An informed decision maker with unique evaluation requirements, not a computerized program or an evaluation technique, makes the investment choices.

Another facet of project evaluation that evaluators often overlook is the discrepancy between the low degree of accuracy with which they can obtain forecasts of prices, cost, and quantities and the higher degree of accuracy possible in computing other cash flow components, such as the discount rate and taxes. It is easy to spend too much time on less important details just because these details can be computed with accuracy. Given the uncertainty in forecasting prices, does it make sense, for example, to use sophisticated depreciation options to calculate taxes in detail or to get the weighted average cost of capital determined to two decimal places? It is important to try to determine each cash flow component accurately, but the limitations inherent in price forecasts over a 20-year period make such a high degree of accuracy for other parameters unnecessary.

Perhaps more important than the choice of evaluation method is the inclusion of all appropriate costs and benefits in the valuation, even though analysts may not know the magnitude of these costs and benefits with accuracy. For example, too little attention has been paid to the shutdown costs of mining projects in past years. These costs can be substantial and must be included in any modern evaluation. In addition, governments and industries should give more thought to social costs. There is now a better appreciation of the importance of these costs and of including them in any large-scale mining plan.

The creation of more sophisticated evaluation methods and the availability of computers to perform the calculations show that project evaluation must make extensive use of computers to be successful. Failure to use the full capabilities of computers will place the project participant at a distinct disadvantage because modern project evaluation relies heavily on computer modeling.

New evaluation methods are continually being developed in an attempt to solve or avoid the many problems associated with discounted cash flow analysis. Option valuation and simulation may solve some of the theoretical shortcomings of DCF analysis under conditions of uncertainty. Monte Carlo simulation, when used in conjunction with investor preference theory, shows particularly high promise. Simulations that include provisions for production decisions may also prove to be powerful evaluation tools. These methods may become more popular in the future.

Only by carefully choosing appropriate evaluation methods, as well as the categories and values of benefits and costs, can the analyst provide sufficient information to the decision maker. Although these methods have faults and shortcomings, they are the best available at present. It is far better to use existing evaluation methods to obtain information on a project's worth than to rely on a subjective opinion.

APPENDIX A

A Review of Discounting and Compounding

The purpose of this appendix is to provide a quick review of or reference for basic functions involving compounding, discounting, and cash flow analysis. This review is not meant to be extensive or all inclusive, but it is meant to illustrate the principles in very simple terms. The review of each topic generally consists of a simple problem that is solved in more than one way to illustrate the principle involved; it is sometimes accompanied by a brief discussion. Further information can be found in any engineering economy textbook.

The topics discussed in this review include the following:

- future value (FV), also known as future worth (FW)
- nominal and effective interest
- continuous compounding
- present value (PV), also known as present worth (PW)
- cash flow (CF) and discounted cash flow (DCF)
- net present value (NPV)
- annuities
- internal rate of return (IRR) and growth rate of return (GRR)
- discount rate and interest rate
- weighted average cost of capital
- constant versus current dollars
- effects of inflation
- use of multiple discount rates
- NPV–IRR ranking conflict
- incremental IRR

FUTURE VALUE

Consider the following problem: If $10 is invested and receives an interest rate of 12% per year, what will the FV be at the end of the third year?

Determining FV involves compounding. The following is a review of steps that we may take to perform future value calculations. (Note that we have a time period that runs from 0 to 3 years.)

Using simple compounding, we obtain the following results:

Year	Calculation and resulting value
0	$10
1	$10 × 1.12 = $11.20
2	$11.20 × 1.12 = $12.54
3	$12.54 × 1.12 = $14.05

Thus, we arrive at a future value of $14.05.

Another approach is to identify a compound factor based on the interest rate and time:

$$1.12 \times 1.12 \times 1.12 = (1.12)^3 = 1.405$$

and multiply by the principal, which leads to

$$FV = \$10 \times 1.405 = \$14.05$$

In more general terms,

$$FV = PV(1 + i)^t \qquad \textbf{(EQ A.1)}$$

where

i = interest rate

t = number of years

which, for this example, yields

$$FV = \$10(1 + 0.12)^3 = \$14.05$$

Given i and t, another option is to look in a compound interest table to determine $(1 + i)^t$. For $i = 0.12$ and $t = 3$, such tables give $(1 + i)^t = 1.405$, which again yields FV = $14.05.

NOMINAL AND EFFECTIVE INTEREST

We can also perform future calculations for periods less than yearly. Suppose we wish to divide a year into n equal periods. Instead of an annual interest rate, we now have an interest rate for the shorter period. This introduces two terms: (1) the nominal interest rate per year, r, is the annual interest rate without compounding; and (2) the effective interest rate per year, i, is the annual rate with compounding. For example, if a bank pays 1.5% interest per quarter, $r = 1.5\% \times 4 = 6.0\%$. If \$1.00 is deposited in the bank for four periods at a nominal interest rate of 6.0%, at the end of the year the account will have grown to \$1.061. In this case, $i = [(\$1.061/\$1.00) - 1] \times 100 = 6.1\%$. In more general terms, the effective interest rate per year is expressed as follows:

$$i = (1 + r/n)^n - 1$$

where n is the number of compounding periods in a year.

Since there are still t years over which compounding can take place, Equation A.1 can be rewritten to determine the future values for compounding periods less than yearly:

$$FV = PV\left(1 + \frac{r}{n}\right)^{nt} \qquad \textbf{(EQ A.2)}$$

where

t = number of years

n = number of periods in a year

r = nominal interest rate

CONTINUOUS COMPOUNDING

As n approaches infinity, $(1 + r)^{nt}$ approaches e^{rt}, the term for continuous compounding. Therefore, at $n = \infty$, Equation A.2 becomes

$$FV = PVe^{rt} \qquad \textbf{(EQ A.3)}$$

which is the expression for continuous compounding. The following explanation is the mathematical proof:

$$FV = PV\left(1 + \frac{r}{n}\right)^{nt}$$

Multiply nt by r/r to get

$$FV = PV\left[\left(1 + \frac{r}{n}\right)^{(n/r)}\right]^{rt}$$

Let $m = n/r$ and rewrite:

$$\text{FV} = \text{PV}\left[\left(1 + \frac{1}{m}\right)^m\right]^{rt}$$

Then $e = \lim(1 + 1/m)^m = 2.718$, as $m \to \infty$, giving

$$\text{FV} = \text{PV}e^{rt}\text{, the expression for continuous compounding}$$

PRESENT VALUE

Most problems in mineral project evaluation involve determining the present value of a future cash stream or cash flow.

Consider this problem: What is a return of $10 worth today if it will not be received until the end of the third year, assuming there is an alternative option of placing available funds in a project that will yield a rate of 12% per year compounded annually?

Determining PV involves discounting, which is the inverse of compounding. We start with the end result ($10 three years from now) and work back to the present, discounting 12% per year:

Year	Calculation and resulting value
3	$10/1.12 = $8.93
2	$8.93/1.12 = $7.97
1	$7.97/1.12 = $7.12
0	$7.12

Alternatively, we could identify a discount factor based on the discount rate and time, as in the determination of future value:

$$1.12 \times 1.12 \times 1.12 = (1.12)^3 = 1.405$$

but this time divide into the principal:

$$\text{PV} = \$10/1.405 = \$7.12$$

In more general terms,

$$\text{PV} = \text{FV}/(1 + i)^t = \text{FV}(1 + i)^{-t} \quad \textbf{(EQ A.4)}$$

Where

i = discount rate

t = number of years

which, for this example, yields

$$\text{PV} = \$10/(1.12)^3 = \$10(1.12)^3 = \$7.12$$

We can also look in a discount table or PV table to determine $(1 + i)^{-t}$ (or use a compound or FV table to determine $[1 + i]^{t}$) for use in Equation A.4. Yet another approach is to use the financial keys on a handheld calculator. Entering $n = 3$, $i = 12\%$, and FV = \$10 will yield PV = \$7.12.

For continuous discounting, use the following formula:

$$PV = FV(e^{-rt}) \quad \text{(EQ A.5)}$$

CASH FLOW AND DISCOUNTED CASH FLOW

A cash flow is the net stream of yearly investments, costs, and revenues. Think of all investments and costs as negative and all revenues as positive. Table A.1 shows a simple 3-year cash flow with investments made in year 0. Discounting the CF at an appropriate discount rate i gives a discounted cash flow. DCF analysis is one of the major tools used to analyze project feasibility (examples are given later in this appendix).

NET PRESENT VALUE

NPV is the present value of yearly income less the present value of all yearly costs, including investments.

Consider Table A.1 from the previous section, and assume a discount rate of 12%. To calculate NPV, we can find the present value of each year's costs and benefits by discounting those values by 12% per year.

$$\text{PV revenues} = 100/(1.12)^1 + 150/(1.12)^2 + 200/(1.12)^3 = \$351.22$$

$$\text{PV costs} = 50/(1.12)^1 + 75/(1.12)^2 + 150/(1.12)^3 = \$211.20$$

$$\text{PV investments} = 125/(1.12)^0 + 15/(1.12)^2 = \$136.96$$

TABLE A.1 Simple 3-year cash flow example

	Year			
	0	1	2	3
Revenue, $		100	150	200
Costs, $		–50	–75	–150
Profit, $		50	75	50
Investment, $	–125	0	–15	0
Cash flow, $	–125	50	60	50

For these present values, NPV is as follows:

$$\begin{aligned}\text{NPV at 12\% discount rate} &= (\text{PV of benefits}) - \text{PV of costs}\\ &= \$351.21 + (\$211.20 + \$136.96)\\ &= \$3.06\end{aligned}$$

Given the same cash flow table, we could also determine the NPV by simply using the yearly cash flow values themselves (making sure to include year 0) and the discount rate of 12%:

$$\begin{aligned}\text{PV CF} &= (-125)/(1.12)^0 + 50/(1.12)^1 + 60/(1.12)^2 + 50/(1.12)^3\\ &= -125 + 44.64 + 47.83 + 35.59\\ &= \$3.06\end{aligned}$$

In more general terms, the formula for NPV is

$$\text{NPV} = \left[\sum_{t=1}^{n} \frac{\text{CF}_t}{(1+i)^t}\right] - I_0 \qquad \textbf{(EQ A.6)}$$

where

CF_t = cash flow in year t

I_0 = initial investment

i = discount rate

n = total number of years

ANNUITIES

An annuity is simply a special form of cash flow in which a constant amount of money is invested in each time period at a constant compound interest rate. We could use PV calculations to determine annuity values, but using the derived annuity formula and yearly factor values allows us to save work. For example, if we have a project that earns $100 per year for 3 years, we could determine the PV as follows (assuming a discount rate, *i*, of 12%):

$$\text{PV} = 100/(1.12)^1 + 100/(1.12)^2 + 100/(1.12)^3 = \$240.18$$

However, to simplify calculations, it can be proven that the PV of an annuity is

$$\text{PV} = \text{PMT}\frac{1-(1+i)^{-t}}{i} \qquad \textbf{(EQ A.7)}$$

where PMT is the annual equal payment made over time period t. Therefore, for $t = 3$ and $i = 0.12$,

$$\text{PV} = \$100\frac{1-(1.12)^{-3}}{0.12} = \$100 \times 2.402 = 240.18$$

As with PV and FV, we can also determine factors for PV of annuities by checking tables or using a suitable handheld calculator.

Familiarity with annuities allows quick and dirty evaluations of cash flows, as well as a method of calculating annualized capital costs. For example, if a copper mine costs $200 million, mine life is 20 years, output is 18,000,000 kg (40,000,000 lb) copper per year, and operating costs are $1.10/kg ($0.50/lb) copper, what price is necessary to achieve a 7% return? In this case, we regard the purchase price of the mine as the present value; thus, PV = $200 million. We also have $t = 20$ and $i = 0.07$, and we can enter these values into Equation A.7 to find that PMT = $18.88 million/year. This is the annual amount of positive yearly cash flow required to recoup all capital in 20 years plus a 7% rate of return on invested capital. Since we know the required annual cash flow and the annual copper output, we can determine the net gain per unit weight of copper:

net gain per unit weight = PMT/annual output
= ($18,800,000)/(18,000,000 kg), or ($18,880,000)/(40,000,000 lb)
= (1.04/kg ($0.47/lb) required for capital recoupment and 7% return

The required price for the copper is then the sum of the operating cost and the capital cost, or $1.04 + $1.10 = $2.14/kg ($0.47 + $0.50 = $0.97/lb).

INTERNAL RATE OF RETURN

IRR is that discount rate that is required to make NPV = 0. The higher the IRR, the more profitable the project, all else being equal. IRR values are frequently used to rank projects, with projects that have the highest IRR being the most desirable. There are problems associated with using IRR as an evaluation criterion, as described later. But first, this section describes how to calculate IRR.

The general formula simply involves setting NPV equal to zero and substituting IRR for i:

$$\mathrm{NPV} = 0 = \left[\sum_{t=1}^{n} \frac{\mathrm{CF}_t}{(1+\mathrm{IRR})^t}\right] - I_0 \qquad \textbf{(EQ A.8)}$$

where

CF_t = cash flow for year t

I_0 = initial investment

n = total number of years

As an example, consider determining IRR for the cash flow information shown in Table A.1. Since we cannot mathematically solve Equation A.8 for IRR, we will solve by trial and error. From the preceding section, we already know that, for a discount rate of 12%, NPV = $3.06. We can calculate successive

NPVs at higher discount rates until we can estimate the required i to make NPV = 0:

i	NPV
12%	3.06
13%	0.89
14%	–1.22

We now know that NPV = 0 for an i value somewhere between 13 and 14%. Although NPV = $f(i)$ is curvilinear, as shown in Figure A.1, we can interpolate to estimate the actual value of IRR.

$$\frac{0.89}{0.89 + 1.22} = 0.422$$

$$13\% + 0.422\% = 13.42\%$$

We can now check to see how close we are to the true IRR by calculating what NPV would be for i = 13.42%. In this case, we find that NPV = –0.01, so the IRR is indeed very close to 13.42%.

Another way to determine IRR is to plot NPV against discount rates, as shown in Figure A.1). For this example, we find that as the discount rate increases, NPV becomes smaller. IRR is that rate indicated when NPV = 0 on the graph.

Another option is to use the financial keys on suitable handheld calculators to determine IRR. Calculators use similar trial-and-error and interpolation algorithms as described earlier.

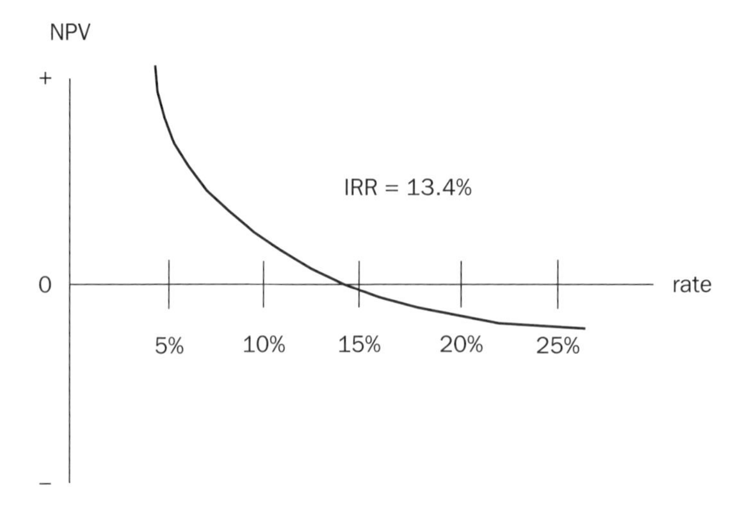

FIGURE A.1 Using a plot of NPV versus discount rate to estimate IRR for cash flow from Table A.1

Multiple-Root Problem

As discussed in the main text of this book, the multiple root problem—in which there may be more than one IRR for a single cash flow—is of concern in IRR analysis. This problem may occur with projects in which large capital expenditures are required both at start-up and in subsequent years. Table A.2 shows two cash flows, CF_A and CF_B, for which there is more than one internal rate of return. Table A.3 shows the NPVs of these two cash flows at various discount rates. The NPV results are graphed in Figure A.2. Each instance where the plot crosses the x-axis corresponds to a zero NPV and, hence, a value of IRR.

There are three IRR values for CF_A and CF_B (the third IRR for CF_B is higher than 40% and is off the scale). The problem is, which IRR to choose? It is tempting to choose, as a rule of thumb, the IRR that is closest in value to the minimum acceptable rate of return (MARR). However, there is no theoretical reason to do so. The other IRR values are also valid.

Though the multiple-root problem is sometimes regarded as a problem solely for IRR analysis, Table A.3 and Figure A.2 demonstrate that NPV is affected as well. As shown in Figure A.2, for certain discount rate ranges, it is possible to achieve a higher calculated NPV value simply by choosing a higher discount rate; for example, the NPV of project B steadily increases across the discount

TABLE A.2 Cash flows for two projects

Year, t	CF_A	CF_B
0	–180	–100
1	100	75
2	100	75
3	100	75
4	100	75
5	100	75
6	–100	75
7	–100	75
8	–100	–200
9	–100	–200
10	–100	–200
11	0	–200
12	0	–200
13	0	50
14	0	50
15	0	50
16	0	50
17	0	50
18	0	50
19	0	50
20	200	300

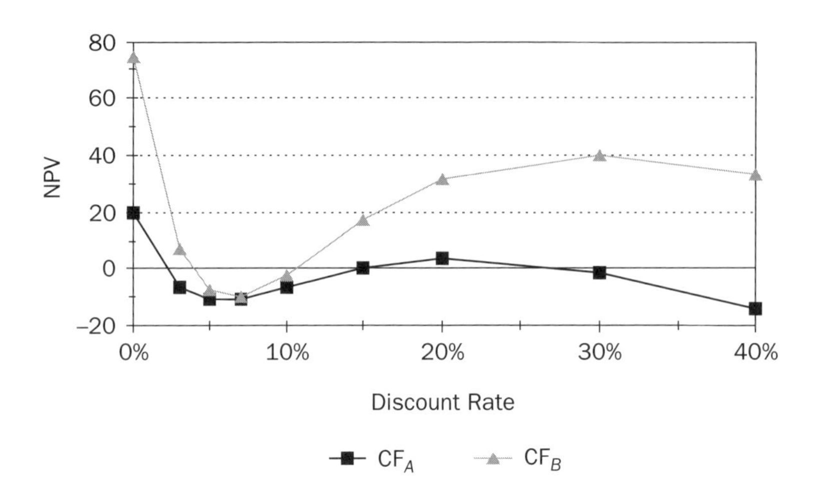

FIGURE A.2 Multiple-root effects on IRR and NPV for projects A and B

TABLE A.3 Net present values for example projects A and B at various discount rates

i (%)	NPV_A	NPV_B
0	20.0	75.0
3	–6.3	7.1
5	–10.9	–7.2
7	–10.6	–9.3
10	–6.6	–1.8
15	0.8	17.2
20	4.1	31.5
30	–1.0	40.1
40	–14.1	33.5
50	–29.1	21.7

rate range of 10 to 30%. In practice an evaluator chooses a discount rate and then accepts the resulting NPV. Choosing the IRR that is closest to the MARR when multiple IRRs are present is consistent with this approach. Since there is no theoretical reason to choose one of the multiple IRR values over the others, there is no theoretical reason to choose the NPV derived from the discount rate that is closest to the MARR.

We can also show that the benefit:cost (BC) ratio, another merit measure, will be affected by multiple roots. The BC ratio is defined as the ratio of the present value of costs to the present value of benefits. It is another means of expressing the rate of return and has many of the same characteristics as IRR. As an example, consider determining the present values of the costs and benefits of project A (see Table A.2) at various discount rates (Table A.4). Figure A.3 shows all these values plotted versus discount rate. The decision

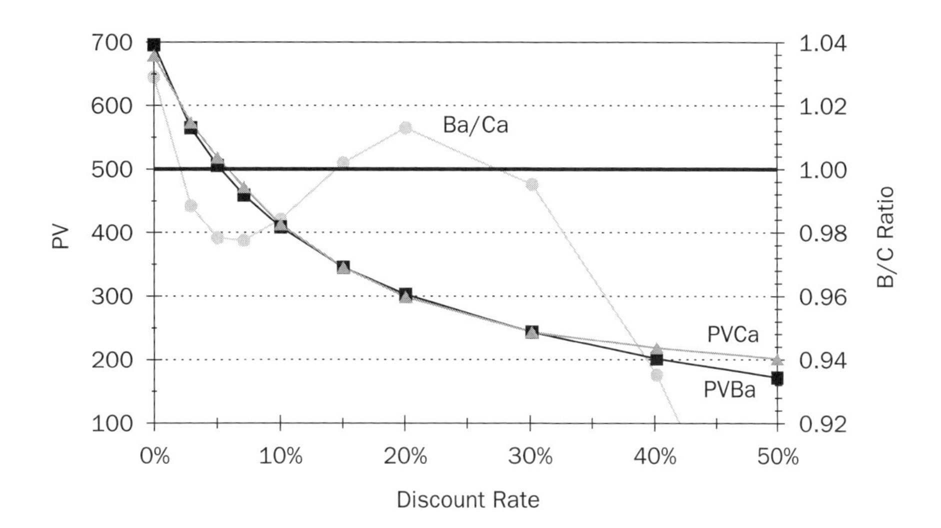

Note: Ba = benefits of project A; Ca = costs of project A; PVBa = present value benefits of project A; PVCa = present value costs of project A.

FIGURE A.3 PV benefits, PV costs, and BC ratio for project A at multiple discount rates

TABLE A.4 PV of benefits and costs and BC ratios at various discount rates

Discount Rate (%)	PV of Benefits ($)	PV of Costs ($)	BC Ratio
0	700.00	680.00	1.03
3	568.71	575.05	0.99
5	508.33	519.23	0.98
7	461.70	472.34	0.98
10	408.81	415.38	0.98
15	347.44	346.66	1.00
20	304.28	300.19	1.01
30	244.61	245.60	1.00
40	203.76	217.84	0.94
50	173.72	202.87	0.86

criterion for BC analysis is that a project is worthwhile if its BC ratio is greater than 1. As the figure shows, there are three discount rate data points for which BC > 1, corresponding to the same three discount rates shown in Figure A.2. This demonstrates that BC analysis has the same multiple-root problem as does IRR.

The end result is that NPV, IRR, and BC ratio—as well as any other DCF-related merit measures—suffer the same problem of values being dependent upon the discount rate when multiple roots exist.

Reinvestment Controversy

There is considerable debate in the literature as to whether IRR analysis requires an assumption that yearly dividends from a project will be reinvested at the IRR. As shown earlier in this appendix, the value of IRR is determined based solely on the number and amounts of the yearly cash flows, as well as the initial investment, given NPV = 0. No reinvestment of yearly dividends is explicitly factored into the IRR value itself. The question is, under what conditions can we view the IRR as being analogous to a compound rate of interest applied to the initial investment amount deposited in a bank account? In other words, under what conditions will the future value of the initial investment amount (if it earns compound interest at the IRR) equal the combined future value for all yearly cash flows from the project?

For example, consider the following project cash flow:

Year	0	1	2	3
Cash flow, $	–150	75	75	75

For this example, the initial investment is $150 and payments are made at the end of the year. We can determine the IRR to be 23.4%. Suppose we could deposit the initial $150 into a bank account where the money would accumulate compound interest for 3 years at an interest rate equal to this IRR (Table A.5). At the end of the 3 years, we end up with $281.80; this represents the FV of $150 compounded at 23.4%. The next step is to determine whether the project with the preceding cash flow must reinvest its yearly dividends at an interest rate equal to the IRR in order to have the same FV as found in the bank account example. Table A.6 shows what happens if such reinvestments are indeed made. The table shows an FV of $281.80 at the end of 3 years. This demonstrates that, if we wish to regard the IRR of a project as analogous to a compound rate of interest applied to the initial investment, we must assume that the project's yearly dividends will be reinvested at an interest rate equal to that IRR. Thus, we can consider reinvestment at the IRR to be an implicit characteristic of IRR analysis.

The argument with this implicit assumption is that, if the IRR is a high value, reinvestment at this rate may be unrealistic. Obviously, reinvestment opportunities at very high rates do not usually exist, which would make the specific

TABLE A.5 Cash flow data for bank savings account example

	Year			
	0	1	2	3
Cash flow during year from the account, $	–150	0	0	0
Amount available at beginning of year to earn interest, $		150	185.1	228.4
Interest earned during year at i = 23.4% (= IRR), $		35.1	43.3	54.3
Amount available at end of year, $		185.1	228.4	281.8

value of IRR inconsistent relative to rates of return on invested capital from alternative investment opportunities. Some authors have argued incorrectly that the need for this assumption makes the IRR less useful as a merit measure. However, the IRR of a project is determined directly by the project's own cash flow profile and is independent of any reinvestment activities or rates. It does not need to represent a realistic reinvestment rate in order to serve as a merit measure. The IRR can still be used to provide meaningful comparisons and rankings of investment opportunities regardless of the reinvestment assumption. The decision criterion is simply that an investment is worthwhile if IRR > MARR. It does not matter how large the difference is; it matters only that the difference is positive.

Growth Rate of Return

It is of course possible to reinvest dividends at some external rate i' other than the IRR. This gives rise to the concept of growth rate of return. For example, consider the 3-year investment cash flow from the "IRR Reinvestment Controversy" section of this appendix, but assume dividends are reinvested at 8%; the resulting cash flow data are shown in Table A.7.

The IRR for the cash flow has not changed; it is of course still 23.4%. However, since the yearly dividends are being reinvested at 8% instead of 23.4%, the total future value of the yearly cash flows in this case ($243.50) is less than the value listed in Table A.6 ($281.80). The growth rate of return in this case is simply the interest rate at which the initial investment amount ($150) would have to be compounded in order to yield $234.50 as the future value at the end of 3 years. Like IRR, GRR is determined by trial and error. In this case, we can find the GRR to be 17.5%.

TABLE A.6 Cash flow data assuming reinvestment of yearly dividends at IRR

	Year			
	0	1	2	3
Cash flow during year from project, $	–150	75	75	75
Amount available at beginning of year to earn interest, $		0	75	167.5
Interest earned during year at i = 23.4% (= IRR), $		0	17.6	39.2
Amount available at end of year, $		75	167.6	281.8

TABLE A.7 Cash flow data for 3-year investment example, assuming reinvestment at 8%

	Year			
	0	1	2	3
Cash flow during year from project, $	–150	75	75	75
Amount available at beginning of year to earn interest, $		0	75	156
Interest earned during year at i' = 8%, $		0	6	12.5
Amount available at end of year, $		75	156	243.5

IRR can be viewed as simply a special type of GRR. Note that reinvestment of dividends is explicitly accounted for in the definition of GRR, whereas it can be considered implicit in the definition of IRR.

DISCOUNT RATE VERSUS INTEREST RATE

The discount rate is the rate used to discount future cash flows to determine present value; it reflects the opportunity cost of capital to the firm. An interest rate is the cost of borrowed funds; an interest rate at which a bank lends funds may or may not reflect the opportunity cost of capital for a firm. Interest rates and discount rates may or may not be the same thing. It is incorrect to use the two interchangeably.

Determining the interest rate a bank charges for a particular set of borrowed funds is easy. Any bank would be glad to provide a quote. Determining the appropriate discount rate to evaluate a project being undertaken by a firm is more difficult. Determining the opportunity cost of capital means accounting for such items as the rate of return on alternative uses of capital, interest on borrowed funds, how interest may change given the level of borrowing and the type of project involved, the expected rate of inflation, and (perhaps) the amount of risk involved.

WEIGHTED AVERAGE COST OF CAPITAL

One of the major problems in conducting cash flow analysis is choosing the proper discount rate. One commonly used or considered discount rate is known as the weighted average cost of capital (WACC). This rate recognizes that there is a cost of equity just as there is a cost of debt and that the debt:equity ratios of firms may vary. It also recognizes and accounts for the fact that the risk of purchasing firms' stocks will vary. The WACC is calculated as follows:

$$\text{WACC} = \frac{D}{D+E}K_f + \frac{E}{D+E}K_e \qquad \textbf{(EQ A.9)}$$

where

D = proportion of debt for the firm

E = proportion of equity

K_e = cost of equity

K_f = after-tax cost of debt

K_f = $(1 - t)K_d$ where t is the effective tax rate

K_d = before-tax cost of debt

The cost of equity in the preceding equation may be calculated as

$$K_e = R_f + (R_m - R_f)\beta \qquad \textbf{(EQ A.10)}$$

where

R_f = return on risk-free securities (T-bond rate)

$\bar{R}_m$ = average market return (obtained from a commercial investment survey, available at any brokerage house or most libraries)

β = the systematic risk of the firm, or the ratio of the volatility of the firm's stock price to market price volatility (this can also be obtained from an investment survey)

CONSTANT VERSUS CURRENT DOLLARS

In times of inflation, money that is received now but spent in the future will have less purchasing power. This decline is caused by monetary or fiscal policies or the growth rate of the economy in general. Prices may also change over time in the absence of inflation simply as a result of basic changes in the supply-and-demand relationships of the good in question. More often, however, prices change over time as a result of both inflation and changes in the supply-and-demand relationships.

In order to conduct a meaningful cash flow analysis, we must keep all accounts either (1) in current dollars, with the yearly inflation rates clearly stated, or (2) in constant dollars, with the stated base year as an index. All items in a cash flow must be identified as being in constant or current dollars. Similarly, constant or current discount rates must be used to correctly determine the value of the cash flow. As stated in the main text of this book, the most frequent error made in DCF analysis is to use an inflated discount rate to evaluate a constant dollar cash flow; the next most frequent error is to compare a constant dollar IRR with current dollar interest rates.

Historically, firms have made real rates of return of only 3 to 4% per year or less. However, it is common to see listed book returns and interest rates of 8 to 15%. These higher rates almost always contain some allowance for inflation. All quoted interest rates are nominal (i.e., include inflation). A bank expects the price of loaned money to include a real return plus a return for accepting risk plus any expected inflation. Therefore, a bank rate of interest, rates for bonds (Treasury, municipal, or otherwise), and a firm's cost for borrowing money are always nominal interest rates. It is incorrect to construct a cash flow on a constant dollar basis and then use the firm's cost of borrowed money or any commercial interest rate as the discount rate without first removing the inflation component.

A nominal interest rate has two components: one to represent the real rate of return, the other to account for inflation. While doing so is not strictly correct, we often add the two to arrive at an inflated or nominal rate. For example, if the real rate of return is 3% and inflation is 7%, we might add the two to arrive at a nominal rate of 10%. Although simple addition gives nearly correct answers, the effects of compounding make this method theoretically incorrect. The correct relationship is

$$r = \frac{(1 + r')}{(1 + r'')} - 1 \qquad \text{(EQ A.11)}$$

where

r = real rate, or rate without inflation

r' = nominal or inflated rate

r'' = the rate of inflation

This equation can be rewritten to solve for r':

$$r' = r + r'' + rr''$$

This shows that the difference between results from the proper formula and the simple addition estimate is rr''. This difference is small if inflation rates are low.

As an example, consider the following data:

Year	0	1	2	3
CF, constant $	–100	50.00	100.00	100.00
CF assuming inflation rate of 3% and ignoring the effect of taxes, current $	–100	51.50	106.09	109.27

What is the NPV of CF (constant $) assuming r = 6.00%? Use r = 6.00% as the discount rate. Then NPV = 120.13.

What is the NPV of CF (current $) assuming r = 6.00% and r'' = 3.00%? Calculate r' = 9.18% as the discount rate. Then NPV = 120.13.

We can also calculate IRR values based on the preceding cash flow data, but again we must adjust for inflation. If we use the constant dollar cash flow, we obtain IRR = 55.58%. If we use the current dollar cash flow, we obtain a current dollar IRR = 60.25%. Note that if we use this latter IRR as the nominal rate, we can calculate the real rate as follows:

$$r = \frac{1 + 0.6025}{1 + 0.060} - 1 = 0.5558\text{, or } 55.58\%$$

This rate equals the IRR obtained using the constant dollar cash flow. This highlights that fact that we must *always* compare IRRs on a constant dollar basis or on a consistent current dollar basis. We cannot mix the two to make comparisons.

EFFECTS OF INFLATION

The difference between conducting project evaluation on a constant or current dollar basis can be significant if the cash flow includes taxes, depreciation, depletion, working capital and any other type of charge or cost that is carried over from one year to following years. Inflation results in higher before-tax profits. However, items such as depreciation are not affected by inflation once the capital expenditures have been made; they do not increase over time. This causes the tax base to become higher each year, which causes higher taxes. Paying more taxes decreases the yearly cash flows and results in a lower NPV. In addition, inflation decreases the relative worth of working capital. Therefore, the net effect of inflation on a cash flow is more taxes, a loss of purchasing power of invested working capital, and lower NPVs.

Since including inflation results in a more accurate evaluation, it may seem unclear why constant dollar evaluations would ever be conducted. However, there are a number of good reasons for doing so:

1. Determining the true expected rate of inflation is difficult. For instance, what will the rate of inflation in the United States be in year 2020? However, increasing use of inflation indexed government bonds may reveal market expectations of inflation (She 1995).
2. The numbers in an inflated cash flow can become so large as to lose their meanings in relationship to numbers that are familiar for today's costs. For example, a reasonable price for coal today would be $36.29/tonne ($40/ton) delivered. At 6% inflation, this price would be $87.09/tonne ($96/ton) in 15 years from now even though there has been no change in the real price of coal over this period. Since forecasts are made of real changes in prices as well as inflation rates, the observed price of $87.09/tonne ($96/ton) 15 years from now does not immediately inform the analyst whether the real price of coal has increased, decreased, or remained constant over this time period without further computation. In this respect, the inclusion of inflation makes understanding the internal workings of a cash flow more difficult for the analyst.
3. The construction and analysis of a constant dollar cash flow are easier and quicker.

In spite of these advantages, evaluators feel uneasy evaluating a project using a risk-free real discount rate of 2 to 3% when the loan document that accounts for inflation and bank risk shows 12%.

USE OF MULTIPLE DISCOUNT RATES

Using the net present value method to evaluate long-lived risky projects poses two major problems:

1. Even though corporations or society may desire long-lived projects over shorter ones, the discounted values of later-year cash flows are essentially zero.
2. Using a constant risk-adjusted discount rate to determine NPV assumes that the risk will not vary during the project life, but in reality project risk usually decreases after a certain point in time.

The first problem implies that conventional discounting somehow undervalues long-lived projects. The second implies that a single risk-adjusted discount rate overstates actual risk, which again undervalues long-lived projects. These problems are particularly acute in the mineral industry, where evaluations of risky, long-lived projects are common.

There are a number of proposed solutions to these problems. The first is to use nonconstant discount rates that reflect the decrease in risk over time. This procedure has merit since risk almost always decreases as the riskier early portions of the project are completed. The questions, then, are what discount rates should be used and when the switch to different rates should occur. The point in time when all invested capital plus the required minimum rate of return has been obtained represents one possible switching point for use in a pro forma cash flow. The procedure for identifying the switch point and switching to a lower discount rate to determine NPV is shown in Table A.8. If the risk-adjusted discount rate consists of MARR = 10% plus 5% to account for risk, then r = 15% and the resulting NPV = 6.3. IRR is calculated at the end of every year. At the end of year 4, the calculated IRR is greater than MARR, which indicates that all capital plus MARR has been recovered. At this point, the discount rate r is lowered from 15% to 10%. The PVs of the cash flows are calculated using r and the sum of the PVs = NPV = 14.8. In this example, it is perfectly rational to use a multiple discount rate and to achieve a higher NPV.

It is also possible to use separate discount rates for each component of the cash flow to reflect differences in risk among these components. For example, working capital has a lower risk of nonrecovery than does invested capital for equipment (it is always possible to liquidate inventories). Therefore, there is a rationale for using a lower discount rate to value the present value of working capital as opposed to the capital used for equipment purchase.

NPV–IRR RANKING CONFLICT

As discussed in the main text of this book, IRR analysis may give different rankings than NPV analysis to mutually exclusive investment opportunities. It is possible for a project with the highest NPV to have the lowest return on investment if IRR is not calculated correctly.

An example of the characteristics of NPV and IRR and the change in rankings of project desirability are shown in Table A.9. For two cash flows, three cases are shown: unequal project lives; unequal initial investments; and unequal

TABLE A.8 IRR-based switch point to determine variable discount rate

	Year					
Parameter	0	1	2	3	4	5
Cash flow, $	–100	32	32	32	32	32
IRR for cumulative years, %			–25	–2	11	18
Variable discount rate (r), %	15	15	15	15	10	10
Present value of cash flow at r, $	–100	27.8	24.2	21.0	21.9	19.9

MARR = 10%
Risk premium = 5%
Risk-adjusted rate, r = 15%
NPV (using r = constant 15% for all years) = $6.3
NPV (using r = variable rate) = $14.8

TABLE A.9 Ranking of projects based on NPV, IRR, and BC ratio for unequal investment and/or unequal project life, as well as unequal service life

	Year								
	0	1	2	3	4	5	NPV ($)	IRR (%)	BC Ratio
				Unequal Project Life					
CF1*	–100	50	50	50	50	50	80.24	41.04	1.80
CF2	–100	65	65	65	0	0	56.12	42.57	1.56
Calculation of CF2*									
CF2	–100	65	65	65	0	0	56.12	42.57	1.56
CFr	0	0	0	–65	0	81.54	0.00	12.00	1.00
CF2*	–100	65	65	0	0	81.54	56.12	36.31	1.56
				Unequal Initial Investment					
CF1*	–100	50	50	50	50	50	80.24	41.04	1.80
CF2	–50	35	35	35	35	35	76.17	64.12	2.52
Calculation of CF2*									
CF2	–50	35	35	35	35	35	76.17	64.12	2.52
CFa	–50	0	0	0	0	88.12	0.00	12.00	1.00
CF2*	–100	35	35	35	35	123.12	76.17	33.78	1.76
				Unequal Project Life and Unequal Investment					
CF1*	–100	50	50	50	50	50	80.24	41.04	1.80
CF2	–50	45	45	45	0	0	58.08	72.45	2.16
Calculation of CF2*									
CF2	–50	45	45	45	0	0	58.08	72.45	2.16
CFr	0	0	0	–45	0	56.45	0.00	12.00	1.00
CFa	–50	0	0	0	0	88.12	0.00	12.00	1.00
CF2*	–100	45	45	0	0	144.57	58.08	30.08	1.58
				Unequal Service Life (equipment life for A is 5 years and for B is 1 year)					
A	–50	15	15	15	15	15	4.07	15.24	1.08
B	–10	13	0	0	0	0	1.61	30.00	1.16
		–10	13						
			–10	13					
				–10	13				
					–10	13			
B sum	–10	3	3	3	3	13	6.49	30.00	1.65

Note: MARR = opportunity cost of capital = 12.00%; all cash flow values in dollars

CF1* = first cash flow, with greater project life and/or initial investment

CF2 = second cash flow, without adjustments to account for lesser project life and/or initial investment

CFr = reinvested cash flow

CFa = cash flow from alternative (opportunity) investment

CF2* = second cash flow, with adjustments to account for lesser project life and/or initial investment project lives and unequal initial investments. In addition, the table shows the impact of unequal service life.

All cash flows are discounted at the same minimum acceptable rate of return, which, for simplicity, equals the opportunity cost of capital (OCC). This assumes the OCC remains unchanged regardless of which projects are undertaken. In the first three cases, a comparison of CF1* and CF2 shows that CF1* has the higher NPV but the lower IRR. The BC ratio is greater for CF1* in case of unequal project life but lower for the other two cases. The different rankings given by the merit measures for CF1* and CF2 in Table A.9, however, simply reflect the fact that the CF2 cash flows are not the correct flows for determining IRRs for comparison purposes. In all three of these cases, one or two of the theoretical requirements for calculating and comparing IRR values (as discussed in the main text) have been violated.

The first violation is that all investments to be compared on the basis of IRR must have equal project lives. In practice, this is seldom the case; thus, as shown in Table A.9, what must be compared are the total cash flows that would result from the two investments over the same length of time. Since CF2 ends in year 3, we must account for income that would be received in years 4 and 5 as a result of investing in CF2. For example, the $65 obtained in year 3 could be reinvested at the OCC for an additional 2 years. This gives CF2*, which can then be correctly compared with CF1*. As Table A.9 shows, the NPV and BC ratio for CF2* are the same as for CF2, but the IRRs are different. Once we account for the additional 2 years, the IRR rankings for CF1* versus CF2* are the same as the rankings obtained using NPV. Note that it might be possible to reinvest in a third project for the 2-year period with different cash flows and different OCC. What is important, though, is that the IRR for the shorter project must be determined assuming reinvestment for the remaining time, whatever the reinvestment opportunities may be.

The second violation is that all projects to be compared on the basis of IRR must have equal initial investments. Since two projects rarely have the same initial investments, we have to assume that the difference in initial investments will be applied to a third investment opportunity. For example, in the case of unequal initial investments in Table A.9, CF1* and CF2 require initial investments of $100 and $50, respectively. If we place $100 into CF1* but only $50 into CF2, then we must account for the earnings of the remaining $50 not invested in CF2. It is the combined cash flow of CF2 and the alternative investment that must be compared with CF1*. If the unspent $50 is invested in a 5-year bond at 12.00%, a cash flow of $88.12 will be available at the end of the fifth year. This is then added to CF2 to produce CF2*, which can then properly be compared to CF1*. As Table A.9 shows, the IRR rankings for CF1* versus CF2* are the same as those obtained using NPV. Note that it might be possible to invest the unspent $50 in any number of other possible investments for the 5-year period. The resulting IRR (as well as the NPV) may be greater or less than that for CF2*, depending on the distribution of the cash flows from the alternative investment. Again, what is important is that we must account for investment of the unspent $50 in the two cash flow comparisons, regardless of how the $50 is invested.

Table A.9 also shows the result of having both unequal project lives and unequal initial investments. The same principles just described apply in this case as well. Again, what is important is that, for the project with the shorter life and lesser initial investment, the unspent capital needs to be invested in a separate opportunity and the money obtained during the shorter project must be reinvested for the remaining 2 years.

Finally, Table A.9 shows the impact of unequal service lives. It would be incorrect to compare a piece of equipment that has a service life of 5 years with one that provides the same service but has a service life of 1 year. For instance, as shown in Table A.9, the NPV of 1.61 for a single short-lived piece of equipment is irrelevant. Rather, the NPV of 6.49 for *five* of the units over the 5-year service life is what should be compared with the NPV of 4.07 for the long-lived piece of equipment. This example illustrates the necessity of comparing projects that provide identical services on the basis of their service lives rather than their project lives.

DETERMINATION OF INCREMENTAL IRR

Numerous references throughout the text have been made concerning the determination of incremental IRR and its use to achieve the same ranking of projects as given by NPV. The following example shows the steps required in calculating incremental IRR and selecting the most favorable project on this basis.

The capital requirements (I_0) and equal annual benefits (PMT) for each of five mutually exclusive projects are shown in Table A.10. All projects last for 20 years, and MARR = 5%. The NPV of each project equals the PV of the benefits less I_0. The cash flows for calculating IRR for each project are also shown in Table A.10. The rankings of the five projects are different when based on NPV as opposed to IRR.

Table A.11 shows the rearrangement of projects in order of ascending values of I_0. This arrangement determines the order of the step-wise comparisons that are successively made between pairs of the projects to determine incremental IRR.

Table A.12 shows the procedure of determining incremental I_0, incremental PMT, and incremental IRR. The procedure consists of choosing two projects, one a defender and the other a challenger, to determine a winner based on whether the incremental IRR is greater than or less than MARR. The process starts by defining the project with the lowest I_0 to be the defender and the project with the next highest I_0 to be the challenger. The incremental (Δ) values are determined by subtracting the challenger values from the defender values. For example, ΔI_0 for challenger B and defender D is 2,420 – 990 = 1,430. ΔPMT is calculated similarly. ΔIRR is determined by creating a new cash flow that consists of subtracting the cash flow for project D from that for project B.

The incremental IRR for increment B-D is 21.1%, which is greater than MARR. This means the incremental investment and incremental benefits make project B preferable over project D. This makes D the loser and B the winner in this first round of comparisons. Project B now becomes the defender to the next challenger, which is project A.

The incremental IRR for increment A-B is also greater than MARR, which means project A is preferred over project B. This makes B the loser and A the winner in the second round of comparisons. Project A now becomes the defender to the next challenger, which is project C.

TABLE A.10 Determination of incremental IRR: Input and calculation of NPV and IRR

	Project				
	A	B	C	D	E
Capital I_0, $	4,620	2,420	6,380	990	9,350
PMT, $	704	440	825	132	880
PV benefits, $	8,773	5,483	10,281	1,645	10,967
NPV, $	4,153	3,063	3,901	655	1,617
IRR, %	14.2	17.5	11.5	11.9	7.0
Rank based on NPV	1	3	2	4	5
Rank based on IRR	2	1	3	4	5
Rank based on I_0	3	2	4	1	5
	Year				
Cash flow	**0**	**1**	**2**	**3**	**20**
CF A, $	−4,620	704	704	704	704
CF B, $	−2,420	440	440	440	440
CF C, $	−6,380	825	825	825	825
CF D, $	−990	132	132	132	132
CF E, $	−9,350	880	880	880	880

Note: Life = 20 years; MARR = 5.0%

TABLE A.11 Rearrangement in order of increasing capital cost

	Project				
	D	B	A	C	E
Capital I_0, $	990	2,420	4,620	6,380	9,350
PMT, $	132	440	704	825	880
PV benefits, $	1,645	5,483	8,773	10,281	10,967
NPV, $	655	3,063	4,153	3,901	1,617
IRR, %	11.9	17.5	14.2	11.5	7.0
Rank based on NPV	4	3	1	2	5
Rank based on IRR	4	1	2	3	5
Rank based on I_0	1	2	3	4	5

TABLE A.12 Calculation of incremental values and determination of winner

	Increment			
	B-D	A-B	C-A	E-A
ΔI_0, $	1,430	2,200	1,760	4,730
ΔPMT, $	308	264	121	176
ΔIRR, %	21.1	10.3	3.2	−2.7
Winner	B	A	A	A
Discarded loser	D	B	C	E

	Year				
Cash flow	0	1	2	3	20
CF B − CF D, $	−1,430	308	308	308	308
CF A − CF B, $	−2,200	264	264	264	264
CF C − CF A, $	−1,760	121	121	121	121
CF E − CF A, $	−4,730	176	176	176	176

In this third set of comparisons, the incremental IRR for increment C-A is less than MARR, which means project A is preferred over project C. A similar conclusion is reached in the last comparison, in which the incremental IRR for increment E-A is also less than MARR.

The result of these step-wise comparisons of incremental IRRs is that project A is preferred over all other projects. The same results could have been obtained by simply observing the values of NPV.

While this process of using incremental IRR results in the same ranking as does NPV, it is cumbersome and its meaning is obtuse. As explained previously in this appendix, the same results can be achieved with less trouble and greater insight by adhering to the basic theoretical requirements of cash flow analysis.

APPENDIX B

A Review of Option Valuation

Graham A. Davis

This appendix demonstrates (1) that mine production can be seen as having an option on reserves and (2) that mines can be valued using option-pricing techniques. In particular, it shows that the only way to value certain types of mines is via an option-pricing framework.

MINE PRODUCTION AS AN OPTION ON MINERALS

First, reviewing the concept of an option is in order. Suppose we have been given the opportunity to purchase 1 oz (i.e., 1 troy ounce, or 31 g) of gold for $400 from a third party in exactly 1 year; after that time the option expires. This 1-year "contract" is known as a European call option on an ounce of gold with an exercise or strike price of $400. A European option allows the transaction at only the contract's expiry. Since gold is, at the time of this writing, trading for less than $400/oz, the option is currently "out of the money." However, since the price of gold may rise above $400/oz by our opportunity to transact, which is 1 year from now, our option has some speculative value. Hence, we would not willingly give it up or throw it away. Its value is the present value of the expected payoffs from possibly exercising the option.

For example, suppose we believe there is a 1% chance that gold will be trading at $500/oz in 1 year's time, in which case we would profit $100 by exercising our option and selling the gold received at the market price. Conversely, we believe there is a 99% chance that gold will remain below $400/oz for the rest of the year, in which case we will not benefit from our option. The option is worth the present value of 1% of $100, or about $0.90, to us. When we consider all of the probabilities of the gold price rising to specific prices above $400/oz within the next year and the payoffs to the option holder from each of the possible price outcomes ($401, $402, etc.), we come up with the present value, or price, of the option.

Sophisticated option-pricing techniques take all of these probabilities, payoffs, and timing of payoffs into account. The simplest and most famous option-pricing formula that economists use to value this type of option is called the Black-Scholes formula.

According to this formula, the present value of the call, C, is

$$C = Se^{-\delta T}N(d_1) - Xe^{-rT}N(d_2)$$

where

$$d_1 = \frac{\ln(S/X) + (r - \delta - \sigma^2/2)T}{\sigma\sqrt{T}}$$

and

$$d_2 = d_1 - \sigma\sqrt{T}$$

where

S = the current gold price (\$/oz)

X = the exercise price (\$/oz)

T = the time remaining in the contract (years)

σ = the standard deviation of the change in the gold price in 1 year (% expressed as decimal)

r = the risk-free rate of interest (%/year expressed as decimal)

δ = the convenience yield on gold (%/year expressed as decimal)

$N(x)$= the cumulative probability distribution function for a standardized normal variable taken from tables supplied as appendices in most finance texts

For our gold option, the input values are S = \$350, X = \$400, T = 1 year, σ = 0.20, r = 0.07, and δ = 0.0.

Plugging these values into the Black-Scholes formula tells us that this 1-year call option on gold has a value of \$11.32. In fact, even if the option offered the chance to purchase gold at \$600/oz 1 year in the future, it would still have some small positive value since there is always a chance that gold will rise beyond the option's exercise price within a year. The preceding formula tells us that such an option would be worth \$0.10. Options are always worth something, regardless of their terms.

Mineral production can be seen as providing such an option, and as such we can value mineral properties as real options using the same techniques we use to value financial options like the one just described. Consider a simple example of a developed gold mine that has been in operation for years and now has only 2,000 oz of reserves remaining. Daily production is 2,000 oz; hence, there is only 1 day of production left. The reserves are known with certainty, and there is no chance that more reserves exist. Assume that the mine receives payment for its output and must pay its bills simultaneously at the end of the day based on start-of-day prices. Operating costs are locked in at

\$400/oz through contracts, but the gold price changes daily and is unhedged. The current gold price is \$350/oz.

Before proceeding to demonstrate how to value this mine using option pricing, consider the traditional NPV analysis of the mine. NPV analysis begins with the assumption that production of the remaining reserves will take place immediately with no chance for delay. Receipts in this case would equal $\$350 \times 2{,}000$, and costs would equal $\$400 \times 2{,}000$. With a net cash flow of −\$100,000 at the end of the day, and based on an annual risk-adjusted discount rate of 10%, the NPV of the mine at the start of the day is

$$\begin{aligned} CF(1+i)^{-t} &= (-\$100{,}000)(1+0.10)^{-1/365} \\ &= -\$99{,}973.89 \end{aligned}$$

Based on this analysis, the mine has a negative NPV and zero value and should be shut down now instead of producing the last 2,000 oz of reserves at a loss.

This mine's reserves would, however, have a positive value if sold in the market. The operating costs are high, but the reserves' market value is not negative or zero as the NPV analysis indicates. What the NPV analysis misses is that the mine owners have the option to delay producing that last 2,000 oz until gold prices improve. That is, the owners have a call option on the gold in the reserves, and the reserves must therefore be priced using option-pricing techniques. Note that the \$400 cost of extracting an ounce of gold is the same as the \$400 exercise price in the call option described earlier. What the mine owners receive for their \$400 is an ounce of refined gold, just as we received an ounce of gold whenever we exercised our European call option earlier. Of course, because the option to mine the gold is currently out of the money, the owners would choose not to exercise the option today. In other words, they will not mine the gold until gold prices rise, if ever.

Assume now that the mine owners have the option to produce the gold in 1 year's time. If they choose not to produce at that time, the gold is lost forever. In this case, the mine can be seen as a portfolio of two thousand 1-year European call options on the gold in the reserves with an exercise price of \$400 each. As with all financial options, the value of these real options on the reserves must be positive; even though producing the gold today is uneconomical, there is some small probability of the gold price rising during the year such that it will become profitable to exercise the options and extract the gold 1 year in the future.

Now that the mine valuation problem is established as an option problem, the next step is to price one of these identical 2,000 options. This problem is normally quite complicated, and analytic formulas can seldom be used. However, we have set up this valuation problem in such a way that we can again turn to the Black-Scholes formula. In this example, the time to expiry is 1 year, the risk-free rate is currently 7% per year, the convenience yield on gold is zero, the standard deviation of the change in gold prices is about 0.20, and the exercise price is \$400. After plugging these parameters into the formula, we find that the option to mine each ounce of gold is worth $11.32. This, of course, is the same result obtained earlier. Given that the mine contains 2,000 oz of gold and that the number of options is equal to the number of proved ounces of reserves, the total option value of the mine is $22,640. Because the extraction cost is currently greater than the current mineral price, the optimal management strategy is to defer production until the economics of the property improves. This strategy creates value that can be revealed only by option-pricing techniques. Such pricing techniques demonstrate that no matter how uneconomical current production is, a mineral property still has some speculative value. This makes sense given that uneconomical properties do trade in the market for positive values. This example property would therefore trade today for up to its option value of $22,640. (This figure is within the ballpark that mining analysts would come up with if asked about the market value of these reserves.)

This example shows that a developed mine can never have a negative value as long as there is an option to defer production. In fact, as long as there are no extreme environmental liabilities associated with a property and as long as the option to defer production exists, the property will have some positive value that may be measured by option-pricing techniques. However, even for this simplistic valuation problem, the answer was not easy to obtain. For example, how would someone know that the convenience yield on gold was zero? And where did the number 0.20 for the standard deviation of changes in the gold price come from? Experience with financial markets and more than a casual knowledge of option pricing provide the answers. These are specialized topics that require in-depth training.

An even bigger problem is the fact that any mine with 1 day of production left will have the option not only to produce exactly 1 year from now, as we have modeled the problem, but also to wait to produce and mine at any time in the future. Such an option is known as an American option. The Black-Scholes formula is applicable only to European options and so would not be appropriate in this case. If we assume the mine owner has up to 10 years to mine the remaining reserves, the value of the reserves rises to about $71/oz using American option pricing. Furthermore, we have assumed that the only uncertainty is price uncertainty. To perform option pricing correctly, we should really take into account cost uncertainty, interest rate uncertainty, and reserve uncertainty. Thus, all this appendix can do is identify the option characteristics of mine valuation through simplistic examples and introduce the types of problems that can be approached with an option-pricing framework. The actual characterization and pricing of these options should in most cases be left to finance professionals skilled in option-pricing techniques.

It is useful to note one other aspect of option pricing that can be applied to mine valuation. Consider a developed mine where production is ongoing and currently economical and the calculated NPV is positive. Even in this case, NPV valuation techniques would tend to undervalue the mine because there is more than one option attached to the mine. This appendix has so far discussed only the option to defer extracting the mineral, which in this example is not relevant, as the deferral option exists only in mines that are not in operation.

For an operating mine, there are management options to vary output, expand capacity, stockpile ore, and change cutoff grades. NPV analysis does not capture the value of these options and therefore tends to undervalue the reserves (recall that these operating options can have only a positive value). Fortunately, these operating options are thought to add only several percentage points to the NPV of the property (Davis 1996), and we can be fairly comfortable with our NPV estimate of a mine's value when the mine is operating at a profit.

In summary, NPV analysis tends to undervalue uneconomical mineral properties. NPV analysis that inherently assumes that production is immediate and will continue until the reserves are exhausted fails because of these very assumptions. If we instead view an uneconomical mine as a set of options to extract reserves in the future, the value of the mine becomes the value of these options, which is always positive. Option-pricing techniques–not NPV analysis–must be used to value uneconomical mineral properties. Unfortunately, these methods are complex and require the skills of financial experts. For more information, see Dixit and Pindyck (1994) and Trigeorgis (1996), which contain a few introductory chapters on the option pricing of real assets.

For profitably operating mineral properties where the cost of extracting a unit of reserves is less than the mineral price, NPV valuation is satisfactory but will still somewhat undervalue the mine. A rule of thumb as to the degree of undervaluation associated with such NPV calculations is to add about 8% to the NPV to take into account the various management options associated with production (see Davis 1996, 1997).

MINE DEVELOPMENT AS AN OPTION ON DEVELOPED RESERVES

Consider now the valuation of undeveloped mineral properties. Assume that the reserves on the property have been explored and delineated and are therefore known with certainty, but that they remain to be developed and there is no obligation to spend the development costs immediately. We can view this valuation as an option-pricing problem in which the property owners have the option of paying the development costs (the exercise price) and receiving developed reserves that they may or may not decide to extract immediately. For example, assume that the same 2,000-oz property discussed earlier is undeveloped and has a total development cost of $40,000 in present value terms. Assume also that development is instantaneous. The NPV of the developed mine was shown earlier to be –$99,973. NPV analysis now assumes that this undeveloped property would be developed immediately and operated immediately, giving an NPV of –$99,973 – $40,000 = –$139,973.

When we view this property as containing an option on developed reserves, we reach a different result. First, note that paying the $40,000 development cost is the same as paying the exercise price of an option. Next, note that when we pay the development cost (the exercise price), we get in return an option to extract 2,000 oz of developed reserves, which currently has an option value of $22,640 based on a spot gold price of $350/oz. The option to develop the reserves is therefore currently out of the money, and we would not wish to develop (i.e., exercise the option) now. However, this does not mean that the undeveloped property has no value; unexpired options always have some positive value. In this case, assume the property owners have a 15-year European option to develop the property. We anticipate future changes in gold prices to occur over this time period, possibly on the basis of historical data. Under these conditions, this undeveloped property's option value may be estimated to be about $3,062, or $1.53/oz. This is calculated using $S = 22{,}640$, $X = \$40{,}000$, $\sigma = 0.20$, $\delta = 0.0$, $r = 0.07$, and $T = 5$ in the Black-Scholes formula. Hence, we again find that NPV analysis, when it yields a negative value, undervalues undeveloped mineral properties and that option-pricing techniques must be used.

In addition to cases when the undeveloped property under consideration has a negative NPV, option-pricing techniques can also be important for marginal undeveloped properties, i.e., properties for which the NPV is less than the development cost but greater than zero. The additional value that option pricing uncovers in these instances comes from the option to postpone development until the economics become even more favorable (Davis 1997). Fortunately, NPV analysis can still serve as a base in the valuation of these properties. Empirical work shows that pricing marginal undeveloped mineral properties as options adds value equivalent to about 3% of the present value of the gross revenue stream (Davis 1996). That is, a second rule of thumb is as follows: For any undeveloped property for which the positive NPV is less than the development cost, calculate the NPV and then add about 3% of the present value of the gross revenue stream to get the true option value of the property.

APPENDIX C

A Review of Risk Tolerance and Certainty Equivalence

Michael R. Walls and Thomas F. Torries

Many investors would say that the expected value concept, which weights the financial consequences by their probabilities, adequately takes risk into account. However, to the decision maker risk is a function not just of the probability distribution of reserve outcomes or financial payoffs but also of the magnitude of capital being exposed to the chance of loss. In theory, using the expected value criterion for decision making implies that the decision maker (or the firm) is totally impartial to money and the magnitudes of potential profits and losses.

Consider the two projects shown in Figure C.1. As evident in this example, although the expected value of project A equals that of project B, the expected value concept fails to give adequate weight to the firm's exposure to the chance of a very large financial loss. In its strictest sense, when utilizing the expected value rule, the decision maker should be indifferent between project A and project B. However, most investors would readily concede that the "risk" associated with each of these projects is quite different. Though project B has a reasonably high probability of success (.50), the payoff structure is much less attractive than that of project A. The expected value concept is inadequate in measuring the trade-offs between the potential and uncertain upside gains versus downside losses for individual projects, as well as groups of projects such as those shown in Figure C.1. The traditional measures of valuation–such as NPV, IRR, or expected value–may lead to inappropriate choices about competing risky investments.

PREFERENCE THEORY AND CERTAINTY EQUIVALENCE

The problem with analysis that yields a probabilistic evaluation of investment opportunities is that a decision maker may find it difficult to use the probabilistic information in choosing the most favorable project among a group of alternative investments. While the results of a probabilistic analysis yield much information, a more advanced analysis using preference theory is required to correctly and consistently interpret the probabilistic results. In

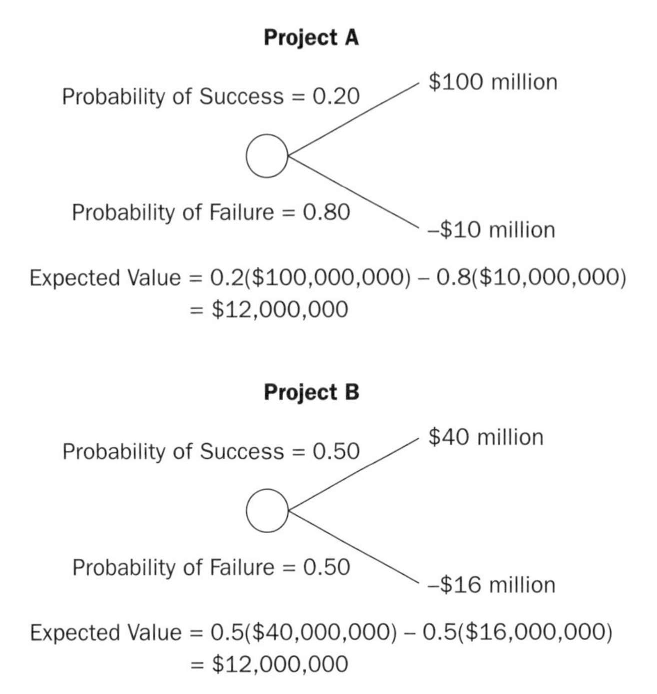

FIGURE C.1 **Projects A and B with equal expected values but different risks**

other words, probabilistic analysis gives results that must be interpreted to be used effectively. The interpretation tool that this appendix discusses is known as certainty equivalence.

Preference theory enables us to construct a utility function that represents the investor's risk attitudes across a range of financial outcomes. One form of utility function that is dominant in both theoretical and applied work in the areas of decision theory and finance is the exponential utility function, which is of the form

$$u(x) = -e^{-x/RT}$$

where

RT = the risk tolerance coefficient

x = the monetary variable

e = the exponential constant

The certainty equivalent, C_x, is equal to the expected value less a risk discount; based on the exponential utility function, the discount is determined by the risk tolerance coefficient, RT, for the investor and the risk characteristics (probability distribution of outcomes) of the investment opportunity. For discrete probability distributions, as shown in Figure C.1, the mathematical expression for the certainty equivalent is

$$C_x = -\mathrm{RT}\ln\left(\sum_{i=1}^{n} p_i e^{-x_i/\mathrm{RT}}\right) \quad \textbf{(EQ C.1)}$$

where

p_i = the probability of outcome i

x_i = the value of outcome i

n = total number of possible outcomes

Consider again the example in Figure C.1, which shows that on an expected value basis the investor should theoretically be indifferent between projects A and B. On a certainty equivalent basis, however, where the investor's attitude about financial risk is incorporated, the investor will favor whichever project has a higher certainty equivalent. For example, for an investor or firm with RT = \$100 million, Equation C.1 shows that project A has a certainty equivalent, C_x, of \$7.2 million while project B has a higher certainty equivalent of \$8.1 million. In this case, the investor would prefer project B over project A. However, another more risk-averse investor with a risk tolerance coefficient of only \$33 million would prefer project A (C_x = \$1.8 million) over project B. (C_x = \$1.4 million). Unlike expected value analysis, the certainty equivalent valuation makes a clear distinction between the "risks" associated with each of these projects. The C_x valuation measures the trade-offs between potential and uncertain upside gains versus downside losses with respect to the investor's risk propensity. In addition, the C_x measure explicitly considers the relative magnitudes of capital being exposed to the chance of loss for each of the projects and the investor's relative strength of preference toward the uncertain financial consequences.

The certainty equivalent valuation also provides guidance to the firm in terms of the value of diversification and risk sharing. Unlike expected value analysis, which provides an "all or nothing" decision rule, the C_x valuation aids the decision maker in selecting the appropriate level of participation consistent with the firm's risk propensity. The participation level represents the percentage of a project the investor undertakes. Other investors jointly participating in the project undertake the remaining percentage. Choosing the proper percentage level is one way to manage risk in the undertaking of a risky project. For example, Figure C.2 shows the certainty equivalent valuation for projects A and B at different participation levels for an investor with a risk tolerance coefficient of \$25 million. For each project, there is an optimum level of participation given the investor's risk tolerance; in this case, both projects have an optimum of approximately 40%. Note also that at levels of participation up to 65%, the investor should prefer project B since it has a greater certainty equivalent than project A. However, for participation levels greater than 65%, project A is the dominant alternative. The important implication in this analysis is that the investor has a formal means of measuring the value of diversification. Note, for example, that the certainty equivalent for either project at the optimum participation level (40%) is considerably greater than the sum of the certainty equivalents for both projects at 100% participation. Also note that participation greater than 90% in project B has a *negative*

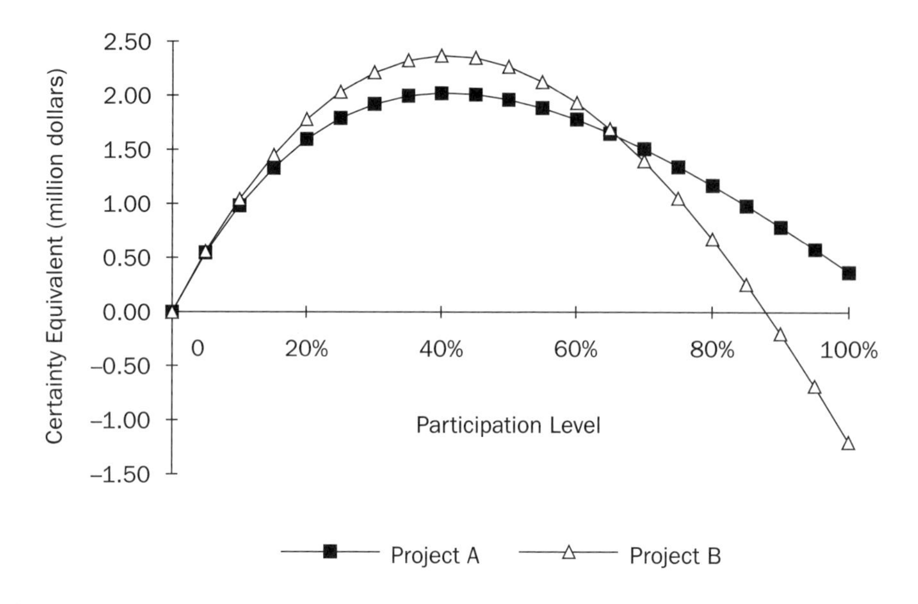

FIGURE C.2 Optimal share analysis on a certainty equivalent basis. This analysis was based on a firm for which RT = $25 million.

certainty equivalent, which implies that this project is too risky for the investor at participation levels greater than 90%.

UNDERSTANDING THE RISK TOLERANCE COEFFICIENT

An advance in the area of preference theory is the use of generic considerations in establishing the risk preferences for an individual or organization. One such consideration is the use of the exponential utility function mentioned earlier to satisfactorily treat a wide range of corporate risk preferences. In the certainty equivalence approach, the risk tolerance coefficient has a considerable effect on the valuation of a risky project. Therefore, it is important to have some intuitive understanding of this measure of risk propensity.

The intuitive notion of risk involves both uncertainty and the magnitudes of the dollar values involved. The central issue associated with measuring an investor's risk tolerance is one of assessing trade-offs between potential upside gains versus downside losses. The investor's attitude about the magnitude of capital being exposed to the chance of loss is an important component of this analysis.

The risk tolerance coefficient is determined based on circumstances in which the investor has an even chance of gaining a certain amount or losing half that amount. Figure C.3 depicts the notion of how RT is determined in terms of decisions about risking money in lotteries with these odds. Lottery 3 offers an even chance of winning $30 million or losing $15 million. An investor who views this investment as too risky would reject it. In other words, the investor's certainty equivalent, C_x, for this lottery would be negative. Lottery 1 offers an even chance to win $20 million or lose $10 million. The investor's

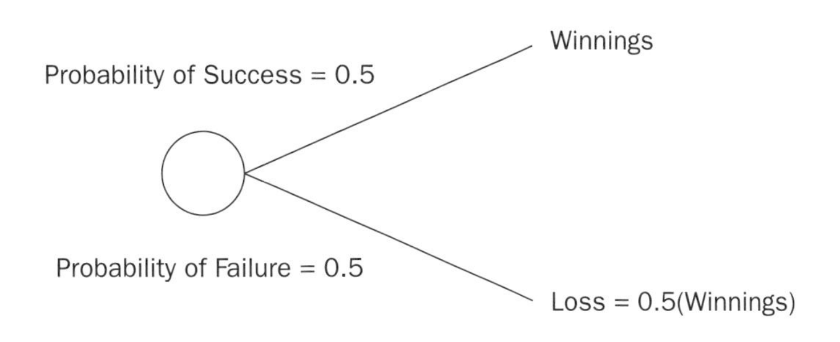

Lottery	Potential Winnings	Potential Loss	Investor Decision
1	\$20 million	\$10 million	Accept ($C_x > 0$)
2	*\$25 million = RT*	\$12.5 million	*Indifferent* ($C_x = 0$)
3	\$30 million	\$15 million	Reject ($C_x < 0$)

FIGURE C.3 Lottery example for determining the risk tolerance coefficient

decision to accept lottery 1 means that the risk-return trade-off associated with this lottery is acceptable given the investor's risk propensity. This type of iterative procedure is continued until we identify the lottery such that the investor is indifferent about entering given an even chance of winning a certain sum or losing half that sum. In this example, that sum is \$25 million and represents the risk tolerance of the investor. Given the investor's indifference, lottery 2 has a certainty equivalent of \$0 for this particular investor.

Relating the risk preferences of the investor to decisions about risky investments, such as in Figure C.3, provides a more meaningful interpretation of the investor's risk propensity. The risk tolerance measure as just described provides meaning to the investor, is reasonably easy to measure, and serves as a sufficiently close approximation to the theoretical risk tolerance. This enables the investor to utilize Equation C.1 for any risky project, compute the firm's certainty equivalent, and make more robust comparisons among risky projects on a cash-equivalent basis.

DETERMINING CERTAINTY EQUIVALENCE FROM PROBABILISTIC DCF

Consider two projects, A and B, with probabilistic distributions of NPV obtained through Monte Carlo simulation (Figure C.4). The most project A may lose is \$40,000, and the most it may gain is \$120,000. Project B has a low probability of significant loss but also a low probability of significant gains; the most it may lose is \$10,000 and the most it may gain is \$40,000. The median value is \$20,000 for either project. The goal is to determine what values an investor would place on the two projects and which project is preferable. The answers depend on the risk preference of the investor.

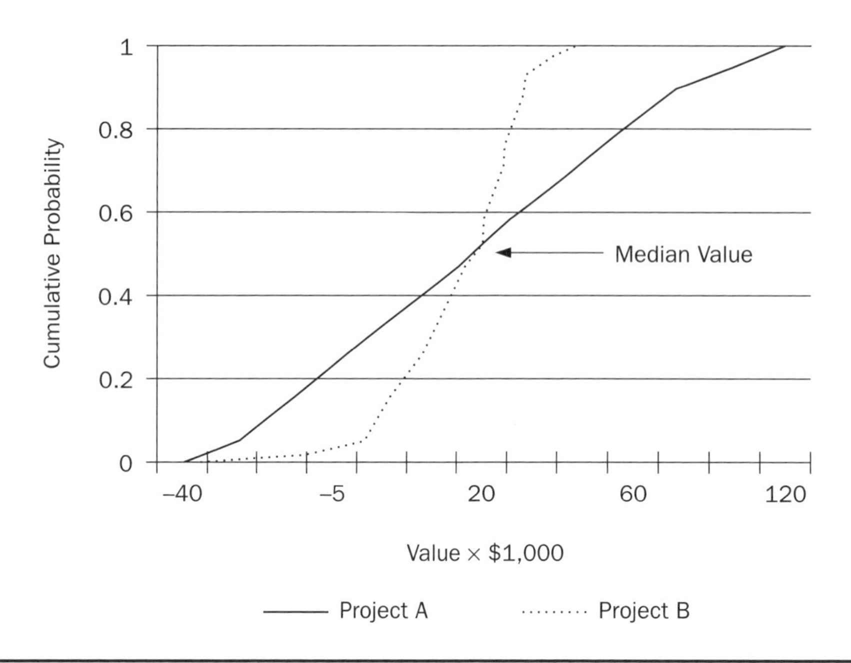

FIGURE C.4 Cumulative probabilistic values for projects A and B

An investor who is averse to the risk associated with losing a given amount of money, regardless of the potential benefits, will reject the project. On the other hand, another investor with greater assets may be glad to undertake the same project. Probabilistic analysis in this case provides the same information to two different investors, who judge the merits of the project differently because of their different perceptions of the risk involved.

Consider the use of the certainty equivalent, C_x, and the risk tolerance coefficient, RT, of an investor to evaluate these two investment opportunities. Figure C.4 shows the probabilistic distribution of values of the outcomes for projects A and B. These distributions must be converted to discrete probabilities so that C_x for projects A and B can be determined for any value of RT. An approach known as the equal probable method is used for this purpose. For this example, the probability distribution (*not* the cumulative distribution) for each project is divided into five intervals; five is usually an appropriate number to estimate the discrete values. Each of the intervals has a single representative point, which is used along with the probability of each interval occurring ($p = 0.20$ for five intervals) to compute the certainty equivalent for each project at different RT values. For a detailed discussion of the equal probable method for determining the discrete values, see Holloway (1976). Table C.1 shows the discrete probabilities of values as determined for both projects. These discrete values and probabilities are then used to calculate the expected values and C_x values for projects A and B at various values of RT. The expected values are $25,000 for project A and $17,600 for project B. The relationships between RT and C_x are shown in Table C.2 and Figure C.5.

Figure C.5 reveals that project B is preferable over project A as long as the investor has a risk tolerance of less than $45,000. Investors with a greater risk tolerance would choose project A. The figure also shows the prices an investor would be willing to pay for the projects. If the investor had a risk tolerance of $20,000, the investor would be willing to pay as much as $15,420 for project B and $9,080 for project A. However, an investor with a risk tolerance of only $5,000 would be willing to pay only $3,080 for project B and nothing for project A.

As mentioned earlier, the expected values of projects A and B, as determined following Monte Carlo simulation, are $25,000 and $17,600, respectively. However, only an investor with a risk tolerance in the range of $200,000–a value quite high given the size of the investments under consideration–would be willing to pay amounts equal to these expected values. This demonstrates that any specific investor may not pay the expected values for the projects. In addition, the expected value for project A is greater than that for project B, which theoretically indicates that project A is more desirable. However, this example has shown that it is not clear that any specific investor would choose project A over project B. Thus, the amount of value an investor places on projects depends not on the expected values but on the probabilistic distribution of values and the investor's attitude toward risk.

TABLE C.1 Discrete approximation of cumulative functions, equal probable method

Project A Interval	Project A Representative Point	Project B Interval	Project B Representative Point	Probability
($40)–($5)	($10)	($20)–$7	$3	.20
($5)–$10	0	$7–$16	$12	.20
$10–$30	$20	$16–$22	$20	.20
$30–$60	$40	$22–$26	$24	.20
$60–$125	$75	$26–$40	$29	.20

Note: All dollar values × 1,000
Expected value, A: $25,000
Expected value, B: $17,600

TABLE C.2 C_x values of projects A and B for various values of RT

RT ($)	C_x for Project A ($)	C_x for Project B ($)
0.05	–9.92	3.08
10	2.56	13.20
20	9.08	15.42
30	12.88	16.15
40	15.29	16.52
50	16.93	16.74
60	18.10	16.89
70	18.98	16.99
80	19.67	17.07
90	20.22	17.13
100	20.66	17.17

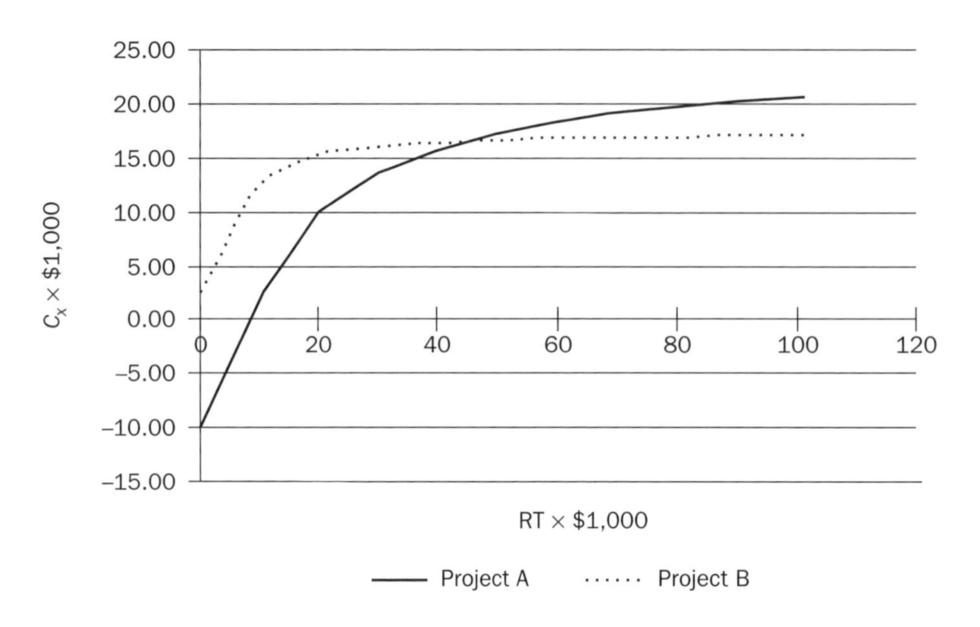

FIGURE C.5 Plot of risk tolerance and certainty equivalence for projects A and B

References

Adams, R. 1991. Managing Cyclical Businesses. *Resources Policy,* June:100–113.

Arizona Department of Revenue. 1994. *Appraisal Manual for Centrally Valued Natural Resource Properties, Tax Year 1994.* Phoenix, Ariz.: Arizona Department of Revenue, Division of Property Valuation and Equalization.

Arrow, K., R. Solow, P. Portney, E. Leamer, R. Radner, and H. Schuman. 1993. *Report of the NOAA Panel on Contingent Valuation.* Washington, D.C.: National Oceanic and Atmospheric Administration.

Au, T., and T. Au. 1992. *Engineering Economics for Capital Investment Analysis,* 2nd ed. Englewood Cliffs, N.J.: Prentice-Hall.

Azzabi, A., and A.R. Khane. 1986. *Manual for Evaluation of Industrial Projects.* Vienna, Austria: United Nations Industrial Development Organization.

Barnes, M. 1980. *Computer-Assisted Mineral Appraisal and Feasibility.* New York: Society of Mining Engineers of the American Institute of Mining, Metallurgical, and Petroleum Engineers.

Baumol, W. 1965. *Economic Theory and Operations Analysis.* 2d ed. Englewood Cliffs, N.J.: Prentice-Hall.

Beaves, R.G. 1988. Net Present Value and Rate of Return: Implicit and Explicit Reinvestment Assumptions. *The Engineering Economist*, 33:275.

Behrens, W., and P.M. Hawranek. 1991. *Manual for the Preparation of Industrial Feasibility Studies.* Vienna, Austria: United Nations Industrial Development Organization.

Bhappu, R., and J. Guzman. 1994. Mineral Investment Decision Making: A Study of Mining Company Practices. Paper presented at the Third Annual Meeting of the Mineral Economics and Management Society, March 24–26 at Washington, D.C.

Bierman, H., and S. Smidt. 1966. *The Capital Budgeting Decision,* 2nd ed. New York: Macmillan Co.

Black, F., and M. Scholes. 1973. The Pricing of Options and Corporate Liabilities. *Journal of Political Economy,* 81(May–June):637–654.

Blackstone, S.L. 1980. Mineral Severance Taxes in Western States: Economic, Legal, and Policy Considerations. *Colorado School of Mines Quarterly,* 75(3):1–33.

Blair, P.D. 1979. Multiobjective Regional Energy Planning: Applications to the Energy Park Concept. In *Studies in Applied Regional Science.* Boston, Mass.: Martinus Nijoff Publishing.

Boyd, J. 1986. *Fundamentals of Coal and Mineral Valuations.* Pittsburgh, Pa.: John T. Boyd Co.

Boyle, H., and G. Schenck. 1985. Investment Analysis: U.S. Oil and Gas Producers Score High in University Survey. *Journal of Petroleum Technology,* April:680–690.

Brennan, M.J. 1993. Martingale Pricing of a Mineral Resource. 1993. Paper presented at the Mineral Economics and Management Society Second Annual Profession Meeting. Feb. 18–20 at Reno, Nev.

Brennan, M.J., and E.S. Schwartz. 1985. Evaluating Natural Resource Investments. *Journal of Business,* 58(2):135–158.

Broadway, R., and F. Flatters. 1993. The Taxation of Natural Resources. Working Paper WPS 1210. Washington, D.C.: The World Bank, Policy Research Department.

Campbell, G. 1990. Diversification or Specialization: The Role of Risk. *Resources Policy,* December:293–306.

Castle, G. 1985. Project Finance—Guidelines for the Commercial Banker. In *Finance for the Minerals Industry.* Edited by C.R. Tinsley, M. Emerson, and W. Eppler. New York: Society of Mining Engineers of the American Institute of Mining, Metallurgical, and Petroleum Engineers.

Cavender, B. 1992. Determination of the Optimum Lifetime of a Mining Project Using Discounted Cash Flow and Option Pricing Techniques. *Mining Engineering,* October:1262–1268.

Chermak, J.M. 1991. Political Risk Analysis: The State of the Art. Working Paper 91-19. Golden, Colo.: Mineral Economics Department, Colorado School of Mines.

A Closer Look at Mining Risk. 1992. *Engineering and Mining Journal,* 193(1):20–21.

Dasgupta, P., A. Sen, and S. Margolin. 1992. *Guidelines for Project Evaluation.* Vienna, Austria: United Nations Industrial Development Organization.

Davis, G.A. 1995a. (Mis)use of Monte Carlo Simulations in NPV Analysis. *Mining Engineering,* 47(1):75–79.

——. 1995b. (Mis)use of Monte Carlo Simulations in NPV Analysis:Discussion. *Mining Engineering,* 47(12):861.

——. 1996. Option Premiums in Mineral Asset Pricing: Are They Important? *Land Economics,* 72(2):167–186.

——. 1997. One Project, Two Discount Rates. Preprint 97-39. Littleton, Colo.: Society for Mining, Metallurgy, and Exploration.

de la Cruz, R.V. 1980. A Critical Examination of the Hoskold Mineral Valuation Method. Preprint 80-3. New York: Society of Mining Engineers of the American Institute of Mining, Metallurgical, and Petroleum Engineers.

Diamond, P., and J. Hausman. 1994. Contingent Valuation: Is Some Number Better Than No Number? *Journal of Economic Perspectives,* 8(4):45–46.

Dixit, A., and R. Pindyck. 1994. *Investment Under Uncertainty,* Princeton, N.J.: Princeton University Press.

Dougherty, E.L., and J. Sakar. 1993. Current Investment Practices and Procedures: Results of a Survey of U.S. Oil and Gas Producers and Petroleum Consultants. Paper 25824 presented at the 1993 SPE Hydrocarbon Economics and Evaluation Symposium, March 29–30 at Dallas, Texas.

Dran Jr., J.J., and H. McCarl. 1974. A Critical Examination of Mineral Valuation Methods in Current Use. *Mining Engineering,* July:71–75.

Dudly Jr., C.L. 1972. A Note on Reinvestment Assumptions in Choosing Between Net Present Value and Internal Rate of Return. *Journal of Finance,* 27:907.

Duvigneau, C., and R. Prasad. 1984. *Guidelines for Calculating Financial and Economic Rates of Return for DCF Projects.* World Bank Technical Paper 33. Washington, D.C.: World Bank.

Eckert, J., ed. 1990. *Property Appraisal and Assessment Administration.* Chicago, Ill.: The International Association of Assessing Officers.

Evans, G. 1984. An Overview of Techniques for Solving Multiobjective Mathematical Programs. *Management Science,* 1(11):1268–1282.

Fletcher, P.B. 1985. Resource Rent Tax Proposal in Australia. In *Finance for the Minerals Industry.* Edited by C.R. Tinsley, M. Emerson, and W. Eppler. New York: Society of Mining Engineers of the American Institute of Mining, Metallurgical, and Petroleum Engineers.

Foley, P. 1982. State Taxation and Copper Supply in the International Context. *Mining Congress Journal,* September:19–23.

Frohlich, A., P. Hawranek, C. Lettmayr, and J. Pichler. 1994. *Manual for Small Industrial Businesses: Project Design and Appraisal.* Vienna, Austria: United Nations Industrial Development Organization.

Garnaut, R., and A.C. Ross. 1983. *Taxation of Mineral Rents.* New York: Oxford University Press.

Gentry, D.W., and T.J. O'Neil. 1984. *Mine Investment Analysis.* New York: Society of Mining Engineers of the American Institute of Mining, Metallurgical, and Petroleum Engineers.

Gillis, M. 1982. Evolution of Natural Resource Taxation in Developing Countries. *Natural Resources Forum,* 22(2):619–648.

Gittinger, J.P. 1982. *Economic Analysis of Agricultural Projects,* 2nd ed. Published for the Economic Development Institute of the World Bank. Baltimore, Md.: The Johns Hopkins University Press.

Glickman, T., and M. Gough, eds. 1990. *Readings in Risk.* Washington, D.C.: Resources for the Future.

Grant, E.L., W.G. Ireson, and R.S. Leavenworth. 1982. *Principles of Engineering Economy,* 7th ed. New York: John Wiley and Sons.

Guzman, J. 1991. Evaluating Cyclical Projects. *Resources Policy* (Butterworth & Co., London), June:114–123.

Hajdasinski, M.M. 1984. Analysis of Internal Rate of Return as an Investment Evaluation Criterion for the Mineral Industry. In *Proceedings of the 18th Symposium of Application of Computers and Operations Research in the Mineral Industry.* New York: Society of Mining Engineers of the American Institute of Mining, Metallurgical, and Petroleum Engineers.

Hanneman, W. 1994. Valuing the Environment Through Contingent Valuation. *Journal of Economic Perspectives,* 8(4):19–43.

Hirschleifer, J. 1961. The Baysian Approach to Statistical Decision: An Exposition. *The Journal of Business,* 34(4):447–489.

Holloway, C.A. 1976. *Decision Making Under Uncertainty.* Englewood Cliffs, N.J.: Prentice-Hall.

Holloway, C.A. 1979. *Decision Making Under Uncertainty: Models and Choices.* Englewood Cliffs, N.J.: Prentice-Hall.

Howard, R.A. 1968. The Foundations of Decision Analysis. *IEEE Transactions on Systems Science and Cybernetics,* SSC-4(3):211–219.

Howard, R.A. 1988. Decision Analysis: Practice and Promise. *Management Science,* 34(6):122–136.

Humphreys, D. 1996. Comment: New Approaches to Valuation: A Mining Company Perspective. *Resources Policy,* 22(1/2):75–78.

Kaufmann, T.D. 1984. Business Cycles and Feasibility Tests in Mining Ventures. *Mining Engineering,* June:610–612.

Keeney, R.L. and H. Raiffa. 1976. *Decisions With Multiple Objectives: Preferences and Value Tradeoffs.* New York: John Wiley and Sons.

King, K.R. 1971. Baysian Decision Theory and Computer Simulation Applied to Multistage, Sequential Petroleum Exploration. Ph.D. dissertation, The Pennsylvania State University, State College.

Labys, W. 1992. The Changing Severity and Impact of Commodity Price Fluctuations. *Overcoming the Third World Economic Crisis.* Essays in honor of Alfred Maizels. Edited by M. Nissanke. Oxford, England: Oxford University Press.

Lane, K. 1988. *The Economic Definition of Ore.* London, England: Mining Journal Books Ltd.

Lehman, J. 1989. Valuing Oilfield Investments Using Option Pricing Theory. Society of Petroleum Engineers (SPE) Preprint 18923, presented at SPE Hydrocarbon Economics Evaluation Symposium, March 9–10 at Dallas, Texas.

Lesser, J., D. Dobbs, and R. Zerbe Jr. 1997. *Environmental Economics and Policy.* Longman, N.Y.: Addison-Wesley.

Lindenberg, E., and S. Ross. 1981. Tobin's q Ratio and Industrial Organization. *Journal of Business,* 54(1):1–32.

Lipscomb, J. 1986. Coal Valuation: The Sales Comparison Approach. *The Appraisal Journal,* April:225–232.

Little, I.M.D., and J.A. Mirrlees. 1990. Project Appraisal and Planning Twenty Years On. In *Proceedings of the World Bank Annual Conference on Development Economics,* Washington, D.C.: The World Bank.

Lohmann, J.R. 1988. The IRR, NPV and the Fallacy of the Reinvestment Rate Assumptions. *The Engineering Economist,* 33(4):303.

Mason, S.P., and R.C. Merton. 1985. The Role of Contingent Claims Analysis in Corporate Finance. In *Recent Advances in Corporate Finance.* Edited by E.I. Altman and M.G. Subrahmanyam. Homewood, Ill.: Richard D. Irwin.

McKnight, R.T. 1988. Empirical Methodology for the Valuation of Risky Gold Cash Flows. In *Gold Mining 88.* Edited by C.O. Brawner. Littleton, Colo.: Society of Mining Engineers.

Mitchell, R., and R. Carson. 1989. *Using Surveys to Value Public Goods: The Contingent Valuation Method.* Washington, D.C.: Resources for the Future.

Newendorp, P.D. 1975. *Decision Analysis for Petroleum Exploration.* Tulsa, Okla.: PennWell Publishing Co.

Newnan, D.G. 1988. *Engineering Economic Analysis,* 3rd ed. San Jose, Calif.: Engineering Press Inc.

Paddock, J.L., D.R. Siegel, and J.L. Smith. 1988. Option Valuation of Claims on Real Assets: The Case of Offshore Petroleum Leases. *The Quarterly Journal of Economics,* August:479–508.

Palm, S.K., N.D. Pearson, and J.A. Read Jr. 1986. Option Pricing: A New Approach to Mine Valuation. *CIM Bulletin,* May:61–66.

Parks, R.D. 1957. *Examination and Valuation of Mineral Property.* Reading, Mass.: Addison-Wesley.

Peters, W.C. 1978. *Exploration and Mining Geology.* New York: John Wiley and Sons.

Pike, R. 1989. Do Sophisticated Capital Budgeting Approaches Improve Investment Decision-Making Effectiveness? *The Engineering Economist,* 32(2):149–161.

Pindred, R.J. 1995. (Mis)use of Monte Carlo Simulations in NPV Analysis:Discussion. *Mining Engineering,* 47(12):861–862.

Pohl, G., and D. Mihaljek. 1989. Project Evaluation in Practice: Uncertainty at the World Bank. Washington, D.C.: World Bank, Economic Advisory Staff.

Pratt, J.W. 1964. Risk Aversion in the Small and Large. *Econometrica,* 32:122–136.

Radetzki, M.S., and S. Zorn. 1980. *Financing Mining Projects in Developing Countries: A United Nations Study.* London, England: Mining Journal Books.

Ramis, F.J., G.J. Thuesen, and T.J. Barr. 1991. A Dynamic Target-Wealth Criterion for Capital Investment Decisions. *The Engineering Economist,* 36(2):107–126.

Ray, A. 1984. *Cost-Benefit Analysis: Issues and Methodologies.* Published for the World Bank. Baltimore, Md.: The Johns Hopkins University Press.

Renshaw, E. 1957. A Note on the Arithmetic of Capital Budgeting. *The Journal of Business,* 30:193.

Rice, T.R. 1974. The Economics of Decision Making. Ph.D. dissertation, Stanford University.

Ridker, R. 1967. *Economic Costs of Air Pollution.* New York: Praeger.

Ross-Watt, D., and B. Mackenzie. 1979. A Mining Project Evaluation Technique Incorporating the Response of Management to the Resolution of Uncertainty. In *Proceedings of the 16th Symposium of Application of Computers and Operations Research in the Mineral Industry.* Edited by T.J. O'Neil. New York: Society of Mining Engineers of the American Institute of Mining, Metallurgical, and Petroleum Engineers.

Samuelson, P. 1976. Economics of Forestry in an Evolving Society. *Economic Inquiry,* 14(4):446–492.

Sarkarat, S. 1996. Evaluation of Mineral Project Using Simulation and Expert System: A Case for a Gold Mine Evaluation. Ph.D. dissertation, West Virginia University, Morgantown.

Savage, L.J. 1954. *The Foundation of Statistics.* New York: Wiley.

Schenck, G.K. 1984. *Handbook of State and Local Taxation of Solid Minerals.* Pennsylvania State University, State College.

——. 1985. Methods of Investment Analysis for the Mineral Industries. In *Finance for the Minerals Industry.* Edited by C.R. Tinsley, M. Emerson, and W. Eppler. New York: Society of Mining Engineers of the American Institute of Mining, Metallurgical, and Petroleum Engineers.

Schreiber, H. 1985. The Role of the Independent Consultant Firm in Project Financing. In *Finance for the Minerals Industry.* Edited by C.R. Tinsley, M. Emerson, and W. Eppler. New York: Society of Mining Engineers of the American Institute of Mining, Metallurgical, and Petroleum Engineers.

She, P. 1995. Benefits and Limitations of Inflation Indexed Treasury Bonds. *Federal Reserve Board of Kansas City Economic Review.* The Federal Reserve Board of Kansas City, 3rd quarter:41–56.

Siegel, D.R., J.L. Smith, and J.L. Paddock. 1985. Option Valuation of Claims on Real Assets: The Case of Offshore Petroleum Leases. Northwestern University Department of Finance Working Paper 4. Evanston, Ill.: Northwestern University.

Smith, R.C. 1992. PREVAL: Prefeasibility Software Program for Evaluating Mineral Properties. IC 9307. Washington, D.C.: U.S. Bureau of Mines.

Solomon, E. 1956. The Arithmetic of Capital Budgeting Decisions. *The Journal of Business,* 29:124.

Squire, L., and H.G. van der Tak. 1975. *Economic Analysis of Projects.* Published for the World Bank. Baltimore, Md.: The Johns Hopkins University Press.

Steiner, H.M. 1992. *Engineering Economic Principles.* New York: McGraw-Hill.

Stermole, F.J., and J.M. Stermole. 1993. *Economic Evaluation and Investment Decision Methods,* 8th ed. Golden, Colo.: Investment Evaluations Corporation.

Thompson, R.A., and G.J. Thuesen. 1987. Applications of Dynamic Investment Criteria for Capital Budgeting Decisions. *The Engineering Economist,* 33(1):59–86.

Tietenberg, T. 1998. *Environmental Economics and Policy.* 2d edition. New York: Addison-Wesley.

Tinsley, C.R. 1985a. Analysis of Risk Sharing. In *Finance for the Minerals Industry.* Edited by C.R. Tinsley, M. Emerson, and W. Eppler. New York: Society of Mining Engineers of the American Institute of Mining, Metallurgical, and Petroleum Engineers.

——. 1985b. Project Finance Supports and Structuring. *Finance for the Minerals Industry.* Edited by C.R. Tinsley, M. Emerson, and W. Eppler. New York: Society of Mining Engineers of the American Institute of Mining, Metallurgical, and Petroleum Engineers.

Tobin, J. 1978. Monetary Policies and the Economy: The Transmission Mechanism. *Southern Economic Journal,* 37:421–431.

Tobin, J., and W. Brainard. 1977. Asset Markets and the Cost of Capital. In *Economic Progress, Private Values and Public Policies: Essays in Honor of William Fellner.* Edited by B. Belassa and R. Nelson. Amsterdam, Netherlands: North-Holland.

Torries, T., A. Rose, and C. Chen. 1988. Mineral Selection and Economic Development: The Case of Namibia. *Materials and Society,* 12(3 and 4):263–293.

Torries, T. 1988. Competitive Cost Analysis in the Mineral Industries. *Resources Policy* (Butterworth & Co., London, England), September:193–204.

——. 1996. Probabilistic Project Evaluation: Here's the Better Mousetrap but Where's the Crowd? In *Proceedings of the Fifth Conference on the Use of Computers in the Coal Industry.* Edited by S.D. Thompson et al. Morgantown, W.V.: West Virginia University, Department of Mining Engineering.

Trigeorgis, L. 1996. *Real Options: Managerial Flexibility and Strategy in Resource Allocation.* Cambridge, Mass.: MIT Press.

UNIDO (United Nations Industrial Development Organization). 1972. *Guidelines for Project Evaluation, Project Formulation and Evaluation Series, No. 2.* New York: The United Nations.

——. 1978. *Guide to Practical Project Appraisal: Social Benefit-Cost Analysis in Developing Countries.* New York: The United Nations.

von Neuman, J., and O. Morgenstern. 1953. *Theory of Games and Economic Behavior.* 3rd edition. Princeton, N.J.: Princeton University Press.

Walls, M. 1995. Integrating Business Strategy and Capital Allocation: An Application of Multi-Objective Decision Making. *Engineering Economist,* 40(3):247–266.

Walls, M.R., and J.S. Dyer. 1996. Risk Propensity and Firm Performance: A Study of the Petroleum Exploration Industry. *Management Science,* Vol. 42, No. 7:1004–1021.

Walls, M.R., and R.G. Eggert. 1996. Managerial Risk-Taking: A Study of Mining CEOs. *Mining Engineering,* March:61–67.

Glossary

balance sheet – A listing of accounts showing all assets, liabilities, and net worth of a project or corporation at a specific time. Assets less liabilities equals net worth. To the extent that book value represents the true value of a project's assets and to the extent that the effects of inflation are negligible, net worth is one means of determining the value of a project or corporation.

benefit:cost (BC) ratio – The ratio of the value of the benefits of a project to the value of the costs. The ratio of the discounted value of the benefits to the discounted value of the costs is properly used as a merit measure in project evaluation. The decision criterion is to choose the project with the highest BC ratio greater than 1.

book value – The listed accounting value of undepreciated assets, which may or may not equal the actual market value of the assets.

by-products – All other simultaneously produced products when one particular product is chosen as the "main" product on the basis of quantity or value. It is not uncommon for more than one product to be produced in a single operation.

by-product credits – Realized revenues from the sale of by-products. By-product credits are deducted from operating costs to determine net operating costs.

capital – Funds required to initiate and sustain an operation. The two primary sources of capital are debt and equity. Capital expenditures are generally made for items that have long lives.

cash flow (CF) – (1) The net sum of all revenues, costs, taxes, and capital, usually determined on a yearly basis, for a company or project. CF can also be determined including debt repayment. (2) A listing of all the components that make up a cash flow as just defined.

certainty equivalent – The value, known with certainty, for which an investor would be indifferent about swapping that value in exchange for a particular risky project.

CF – *See* cash flow.

compounding – The act of geometrically increasing a value, such as money, over a period of time at a specific rate. $FV = PV(1 + i)^t$ is the generalized formula for compounding a present value (PV) over a time period (t) at a specific rate (i) to determine the compounded or future value (FV).
See also discounting; discount rate.

constant costs – Costs in terms of today's dollar. Constant costs exclude the effects of inflation.
See also current costs.

covenant – An agreement between borrower and lender that restricts actions of the borrower to protect the interests of the lender.

current costs – Costs in terms of actual or real prices. Current costs include the effects of inflation.
See also constant costs.

debt – Capital acquired by borrowing on a contractual basis. Such a contract includes guarantees and provisions for the repayment of the amount borrowed and payment of interest for the privilege of borrowing the funds. Lenders are guaranteed by the contractual agreement that all interest due will be paid and that all borrowed funds will be repaid. Debt instruments include long- and short-term bank loans, bonds, preferred stock, and other promissory notes.

debt:equity ratio – For a specific project or company, the current amount of debt divided by the amount of equity in the project. Projects with high debt:equity ratios are highly leveraged and represent higher risks to lenders than do projects with low debt:equity ratios, all else being equal.

deduction – A legal means to reduce taxes. Deductions are specified by the tax codes of a country and are enacted for particular purposes, such as to encourage economic activity in specific areas of the economy.

depreciation and capital allowance – Deductions that are based on a project's actual capital expenses and that represent noncash expenses for the calculation of taxes. These deductions are prescribed by a country's fiscal codes and are applied to gross profit to reduce taxes paid and to encourage further investment in the country. Depreciation schedules and allowances may or may not reflect the useful lives or value of capital expenditures.

discounting – The act of geometrically decreasing a value, such as money, over a period of time at a specific rate. $PV = FV/(1 + i)^t$ is the generalized

formula for discounting a future value (FV) over a time period (t) at a specific rate (i) to determine the discounted or present value (PV).
See also compounding.

discount rate - That rate used to discount the value of future benefits and costs to its present value (i.e., to account for the fact that an amount of money to be received in the future is worth less than the same amount if received today). While the opportunity cost of capital is theoretically the correct discount rate to use, measuring the opportunity cost of capital is sometimes difficult. Consequently, there are many rates that are commonly used as discount rates, such as the weighted average cost of capital, hurdle rates, social discount rates, and safe and risk-adjusted rates.

economic rent - That amount of profit left after all costs, including invested capital recoupment plus a minimum acceptable return on invested capital, have been deducted from revenues. Economic rent is that portion of profit that can be taxed away without affecting future decisions of investors. Economic rents may vary in the short and long terms. Identification of economic rent is crucial for governments wishing to set appropriate tax policies and extract maximum amounts of benefits or revenues.

effective interest rate - The annual rate of interest with compounding. If $1.00 is deposited in a bank that offers 1.5% interest per quarter for four quarters, the amount will grow to $1.061. The effective interest rate is 6.1%.
See also nominal interest rate.

equity - Capital acquired from previously retained earnings of project investors. There is no guarantee that investors will recover invested equity or that they will make a return on the equity.

exchange rate - The worth of one currency relative to another at a specific time. Exchange rates depend upon the combined effects of fiscal policies, trade balances, and inflation rates of the countries in question.

exit cost - The cost of closing an operation. Common exit costs include reclamation, cost of unemployment and reduction of trade, and write-down of debt.

expenses - Costs of items that are immediately used up during the production process, such as labor, energy, and materials. These costs are usually allowed as deductions from revenues for the purpose of determining taxes.

externalities - Costs or benefits not included in the price of a good or service. Exclusion of these costs or benefits distorts market responses, leads to inefficiencies in the market, and retards development of economies. Common examples of externalities are the unrecognized costs of health, social, and environmental problems related to mineral projects.

fair market value - The value a willing and knowledgeable buyer and a willing and knowledgeable seller would be likely to place on a property in the absence of any circumstances that would obligate or force the owner to sell or the buyer to purchase. Fair market value may or may not be the same as net present value, replacement value, book value, or liquidation value.

FC – *See* fixed costs.

fixed costs (FC) – Those portions of operating costs that do not change with output. Administrative and overhead costs are often considered fixed costs in the short run.

future value (FV) – What a cost or income will be worth at a particular time in the future, given an interest or compounding rate.
See also discounting; compounding.

FV – *See* future value.

growth rate of return (GRR) – That discount rate that makes net present value equal zero and that, when viewed as a compound rate, makes the future value of the investment amount equal the future value of the yearly dividends if those dividends of the project are reinvested at some external rate. GRR is determined from a cash flow (CF) in the same manner as internal rate of return (IRR), but the difference between GRR and IRR is that the GRR CF requires dividends to be externally reinvested at a rate different from the IRR.
See also internal rate of return.

GRR – *See* growth rate of return.

hurdle rate – A discount rate commonly used by corporations to evaluate net present value of projects. This rate is based on the best guess as to an appropriate discount rate considering the volatile and uncertain nature of other, more theoretically correct rates, such as the opportunity cost of capital and the weighted average cost of capital.
See also discount rate.

income statement – A listing of accounts that shows the sources and applications of funds of a project or corporation at a specific time. This statement shows all revenues, operating expenses, interest expenses, taxes, overhead costs, debt repayments, and capital expenditures.

incremental analysis – A mathematical process by which the IRR values of the differences in investments and benefits of two or more projects are used to determine NPV-consistent rankings of project worth.

inflation – An economic event in which increases in the volume of money or credit relative to the amount of goods and services available cause substantial and proportional increases in prices in general.
See also discount rate.

initial capital cost – All after-tax costs required to put a project into operation.

interest – The price or cost of debt; the contractual payment for borrowed funds. Quoted interest rates are always given in current dollar terms and include expected effects of inflation. Interest paid is usually considered to be a current expense and is usually deducted as an operating cost for the calculation of taxes.

internal rate of return (IRR) – That discount rate that makes net present value equal zero and that, when viewed as a compounding rate, makes the future value of the investment amount equal the future value of the yearly dividends if those dividends are reinvested at a rate equal to the IRR. IRR can be used to compare rates of returns from alternative investment opportunities, such as other projects, bonds, and savings rates, or minimum acceptable hurdle rates. IRR can also be used to rank mutually exclusive projects, but the ranking may be different from a net present value ranking unless the IRR is calculated correctly for each project.
See also growth rate of return; overall rate of return; hurdle rate; incremental analysis.

IRR – *See* internal rate of return.

marginal analysis – The method of determining the worth of an investment in which the marginal benefit is compared to the corresponding marginal investment. This is the correct method to determine the worth of an investment.

MARR – *See* minimum acceptable rate of return.

mean – The weighted average of all possible values of a variable, where the weights are the probabilities associated with the variables.

minimum acceptable rate of return (MARR) – A generic term for the lowest rate at which the amount of invested money will increase and still be acceptable to the investor. The source and characteristics of the rate are unspecified.

mutually exclusive projects – Projects that are related in that only one of a set of projects can be undertaken. Marginal analysis is required to evaluate mutually exclusive investments unless net present value (NPV) is used as the investment decision criterion. Under conditions of budgetary constraints, the set of investments that maximize total NPV is sought.

mutually independent projects – Projects that are unrelated and can be evaluated and initiated without consideration of other projects. The choice of projects to be undertaken does not require the application of marginal analysis.

net present value (NPV) – Present value of all benefits less the present value of all costs, including initial capital costs. Any NPV is associated with a discount rate; the higher the discount rate, the lower the NPV. NPV can be used to rank projects whether or not they are mutually exclusive.
See also present value.

net worth – *See* balance sheet.

nominal interest rate – A rate of annual interest without compounding. If a bank pays 1.5% interest per quarter, the nominal interest rate is 6%.
See also effective interest rate.

normative economics – That branch of economics that deals with issues for which there are no precise answers because the issues are based on value judgments and morals.
Contrast with positive economics.

NPV – *See* net present value.

ongoing capital cost – Capital required to replace worn equipment and to keep an existing project in operation.

opportunity cost of capital – The return that could have been made had capital been invested in the next best alternative project. Opportunity cost of capital is the correct choice for use as a discount rate in NPV calculations, but it may be difficult to determine.
See also discount rate.

option – The right—but not the obligation—to buy or sell a good, property, or service for a fixed price. This right has a value called the option price.

ORR – *See* overall rate of return.

overall rate of return (ORR) – The rate of return that is obtained by dividing the difference of the present value of the future cash flows less the present value of the initial investment by the present value of the initial investment. This is an unambiguous determination of the actual experienced rate of return on invested capital. ORR also equals (1–present value ratio).
See also internal rate of return; growth rate of return; present value ratio.

payback – The number of years required for cumulative cash flow (which includes initial investments) to reach zero. Payback is a measure of risk and can be useful in evaluating projects when used in conjunction with other project evaluation methods. Payback is also known as payout.

payout – *See* payback.

positive economics – That branch of economics that deals with measurable factors, such as prices, costs, and quantities.
Contrast with normative economics.

present value (PV) – The value of a cost or income to be received in the future when discounted to the present by the use of a discount rate. PV represents what a rational person would value a future income or cost in today's dollars, provided an appropriate discount rate is used. PV is often equated with price.
See also discounting; compounding.

present value ratio (PVR) – A ratio obtained by dividing the present value of the cash flows by the initial investment.
See also internal rate of return; growth rate of return; benefit:cost ratio; overall rate of return.

probabilistic analysis – An evaluation method by which all probabilities of all combinations of possible changes in factor values, such as costs and prices, are considered to determine the probabilistic values of a project.

PV – *See* present value.

PVR – *See* present value ratio.

ratio – A means of judging performance of a project or of defining a limitation placed on a borrower by a lender to protect the interests of the lender. Two commonly used ratios in debt arrangements are reserve:production and debt:equity ratios.

rent – *See* economic rent.

residual value – *See* scrap value.

return on investment (ROI) – Benefit from an investment as a percentage of capital invested in a project. Benefit, in this case, is defined as the net sum of revenues, costs, taxes, and invested capital. Return on investment can be calculated a number of ways. Major variations involve the time periods considered and whether the time value of money is recognized or ignored. Return on investment is what investors get paid for taking risks.

risk – The possibility or the degree of probability of the occurrence of unanticipated events. Risk is usually associated with the possibility of undesirable events and represents the element in investment for which investors are paid.

risk-adjusted rate – A rate that is adjusted to account for risk. The higher the risk and the higher the required returns, the higher the risk adjustment to a specified rate will be.
See also risk; discount rate.

risk tolerance – A measure of the amount of risk an investor will tolerate given specific levels of investments and returns. A measure of risk tolerance is needed to determine the amount an investor would pay for a risky project or the amount by which the investor would discount the expected value of a project to account for risk.

ROI – *See* return on investment.

ROR – *See* return on investment.

safe rate – The rate that can be obtained from an investment that is considered to be safe or to have a low degree of risk. U.S. Treasury bills are considered safe or low-risk investments, and the rate of return of these bills is

considered a safe rate of return. Safe rates of return can be used as proxies for the opportunity cost of capital in the determination of net present value.
See also discount rate.

scenario analysis – Investigation of the results expected from numerous specific combinations of input factor values for a project.
See also sensitivity analysis.

scrap value – The actual market value of a project at the end of the project's life. This includes the actual value of all equipment, improvements, and land and may or may not equal undepreciated book value. Scrap value is also called residual value.

sensitivity analysis – An evaluation method by which input factor values, such as costs and prices, are changed, individually, to determine how variations in the inputs effect the project's value.
See also scenario analysis.

shadow prices – Prices that reflect true economic values of factors, such as labor or energy costs, in countries where government actions have distorted the actual market prices. The use of shadow prices rather than actual market prices to evaluate a project is more appropriate to measure the true benefits and costs of the project to the country's economy. If distorted market prices were to be used, the choice of projects could be incorrect, which would make the country less well off than it could be otherwise.

social discount rate – The discount rate that should be used by governments to evaluate the effects projects will have on the social and economic aspects of the country. The social discount rate is lower than rates used by private investors to evaluate projects because governments do not incur government risk, such as political risk, as do private investors. There are arguments that the appropriate social discount rate should be very low or zero, since a positive social discount rate means that people value their own present consumption greater than that of future generations.
See also discount rate.

sunk costs – Capital already spent. Sunk costs do not enter into decisions concerning the future.

trap – An evaluation procedure in which the mathematics of the evaluation are correct but in which an incorrect model, incorrect assumptions, or incorrect interpretation of results give faulty conclusions. It is a trap because the insidious nature of the incorrect data, model, or interpretation is not always readily apparent.

variable cost (VC) – Any portion of operating costs that changes with output. Labor and energy are generally considered variable costs in the absence of long-term contractual agreements. The period of time involved also determines variable cost; in the long run, all costs are variable.

VC – *See* variable cost.

WACC – *See* weighted average cost of capital.

weighted average cost of capital (WACC) – A discount rate that reflects the proportional cost of capital and debt for a particular project or corporation at a specific time. WACC can be used as a discount rate in the determination of net present value and can be calculated if sufficient information on the industry or company is available.
See also discount rate.

working capital – Capital required to fund inventories and accounts receivable. Working capital fluctuates in proportion to production costs and sales and is recouped entirely when the operation ceases.

Index

Note: *f* indicates figure, *t* indicates table, *n* indicates footnote.